12.37
Ins
11-98

ELECTRONIC PROJECTS FOR THE 21ST CENTURY

ELECTRONIC PROJECTS FOR THE 21ST CENTURY

WRITTEN BY:
JOHN IOVINE

A Division of Howard W. Sams & Company
A Bell Atlantic Company
Indianapolis, IN

©1997 by Howard W. Sams & Company

PROMPT© Publications is an imprint of Howard W. Sams & Company, A Bell Atlantic Company, 2647 Waterfront Parkway, E. Dr., Indianapolis, IN 46214-2041.

International Standard Book Number: 0-7906-1103-1
Library of Congress Card Catalog Number: 97-65785

Acquisitions Editor: Candace M. Hall
Editor: Loretta L. Leisure
Assistant Editors: Pat Brady, Natalie Harris
Typesetting: Loretta L. Leisure
Cover Design: Christy Pierce
Layout Design: Loretta L. Leisure
Graphics Conversion: Bill Skinner, Terry Varvel, Ryan Tungate
Illustrations and Other Materials: Courtesy of the Author

PRINTED IN THE UNITED STATES OF AMERICA

9 8 7 6 5 4 3 2 1

TABLE
OF CONTENTS

◆

◆

INTRODUCTION

Sometimes people ask where do I get my ideas for projects to write about. My answer is simple, I write what I would like to read. This book contains such science experiments.

The projects in this book touch upon technologies that have far reaching implications in the future, such as the recombinant DNA, genetic evolution and holography projects. Other projects apply old technology to new problems such as waste disposal, conservation, recycling and energy generation illustrated in the bio-gas, coal gasification, alcohol fuel production and wood experiments.

The experiments provide you with a beginning into exciting areas of science. It is left to you, if you choose, to blaze your own trail.

CHAPTER 1

◆

RECOMBINANT DNA EXPERIMENT

Gene manipulation is a relatively new science with a tremendous potential. Bio-Tech companies are developing genetically engineered vaccines that are safer and more effective than conventional produced products. Genetically engineered bacteria have produced human lifesaving proteins like insulin and interferon.

Biotechnology has passed from just using simple bacteria to producing human proteins. British researchers use cows, enhanced with foreign DNA, to produce pharmaceuticals in their milk. The genetically engineered cows will reproduce normally, transmitting the genetic code to their offspring.

Plants that produce fruits and vegetables also benefit from genetic engineering. They are made more resistant to attack from bacteria and insects, in some cases producing their own pesticide.

Tomatoes have been engineered that take twice as long to ripen. As a result, the tomatoes will stay fresh on long trips across the country, from farms to market.

There are new advances in growth hormones that will increase milk production in cows, reduce fat in pigs and bring poultry to market faster.

Mammals are genetically altered to produce human tissue. One photograph that received a lot of press was of a mouse that had a human ear growing on its back. Obviously these advantages when perfected, could insure genetically perfect organs for anyone requiring a transplant.

More recently, scientists have cloned sheep. Are humans clones next? President Clinton ordered a moratorium on any government funded projects, forbidding any research on human cloning, citing moral issues. Moral issues aside, this moratorium only applies to U.S. funded laboratories. Other countries (Japan, Russia, China, to name a few) are actively pursuing this research and will reap the economic and scientific benefits this research promises, such as rejection free transplants, nerve regeneration, increased longevity with reduced degeneration, etc.

HUMAN GENOME PROJECT

It has been estimated that it will cost 3 billion dollars and 15 years to map and identify the 100,000 genes within the human's 46 chromosomes. This is equivalent to mapping approximately 3 billion base pairs of DNA. Gathering this information will help scientists to diagnose and treat inherited disorders. In addition, in will provide inroads to curing diseases such as cancer and AIDS.

BIOTECHNOLOGY 101

Before we get to manipulating genetic information, let's review some basic biology. Protein molecules are made up of subunits called amino acids. Two linked amino acids are called a peptide. Larger amino acid molecules are called polypeptides. A protein is a polypeptide molecule. Amino acids are built from atoms of carbon, hydrogen, oxygen, nitrogen and occasionally sulfur. There are twenty different types of amino acids.

AMINO ACID	ABBREVIATION	AMINO ACID	ABBREVIATION
Glycine	GLY	Arginine	ARG
Alanine	ALA	Asparagine	ASN
Valine	VAL	Glutamine	GLN

AMINO ACID	ABBREVIATION	AMINO ACID	ABBREVIATION
Leucine	LEU	Cystine	CYS
Isoleucine	ILE	Methionine	MET
Serine	SER	Phenylalanine	PHE
Threonine	THR	Tyrosine	TYR
Aspartic Acid	ASP	Tryptophan	TRP
Glutamic Acid	GLU	Histidine	HIS
Lysine	LYS	Proline	PRO

Amino acids can be linked in any order. Any change in the order will create a molecule with its own individual properties. For instance, hemoglobin contains 539 amino acids. If you alter just one amino acid out of its proper order it would severely affect the oxygen carrying capacity.

Enzymes are protein molecules that promote a chemical reaction. Enzyme is the term used to describe any biological catalyst associated with living tissue. Enzymes control all chemical reactions in the cell. Enzymes, like physical catalysts, are unchanged by the reaction they promote and can be reused. Enzymes can take apart or put together molecules.

It is the particular enzymes the cell manufactures that determine what type of cell it is. Whether it's a heart cell, brain cell or liver cell.

GENES

For many years it was believed that proteins communicated the genetic information from generation to generation. Although DNA had been looked at as a possible mechanism, it was dismissed as being to simple in structure to relay the complex biological messages, being composed of just four nucleic acids. Proteins, on the other hand, use 20 different amino acids. So it was felt that the 20 amino acids provided a more diverse and complex chemical language to transfer genetic information.

This turned out to be wrong. The simpler nucleic acids in DNA transmit genetic information.

NUCLEIC ACIDS

There are two kinds of nucleic acids; Deoxyribose Nucleic Acid (DNA) and Ribose Nucleic Acid (RNA). Nucleic acid molecules are made up of simpler units called nucleotides. A nucleotide is made up of three components; sugar, phosphate and a base, see **Figure 1-1**.

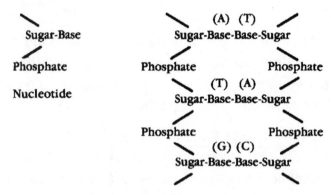

Figure 1. (left) A nucleotide showing a sugar, base, and phosphate as a free molecule and (right) in a DNA ladder.

There is one type of phosphate, two types of sugar, ribose and deoxyribose, (see **Figure 1-2**) and five types of bases.

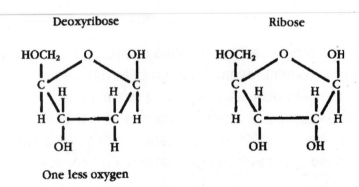

Figure 2. Sugar molecules, (left) deoxyribose and (right) ribose.

BASE	ABBREVIATION	
Adenine	A	
Thymine	T	found exclusively in DNA
Guanine	G	
Cytosine	C	
Uracil	U	found exclusively in RNA

Nucleic acids whether DNA or RNA, are made up of just four nucleotides. In addition, just one type of sugar molecule is used in a particular acid; ribose is used exclusively in RNA and deoxyribose in DNA.

DOUBLE HELIX

In the early 1950s Rosalind Frankin took a sharp x-ray diffraction photograph of the DNA molecule. The diffraction pattern suggested a helical molecule with a repeat pattern of 34 A and a width of 20 A.

James Watson and Francis Crick used this information to work out the structure of the DNA molecule in 1953. The structure of the DNA molecule is a double helix, looks like a twisted ladder or spiral staircase, see **Figure 1-3**. To extended the twisted ladder analogy, the sugar and phosphate of each nucleotide formed the legs, with the bases forming the cross bars.

There is a complementary relationship between the nucleotides. Adenine (A) always pairs with Thymine (T) and Cytosine (C) always pairs with Guanine (G).

The structure of DNA explained how DNA replicated itself during cell division. The hydrogen bonds between the nucleotides break, unzipping the DNA ladder. Each complementary half serves as a template for reconstructing the other half. The results are two identical DNA molecules.

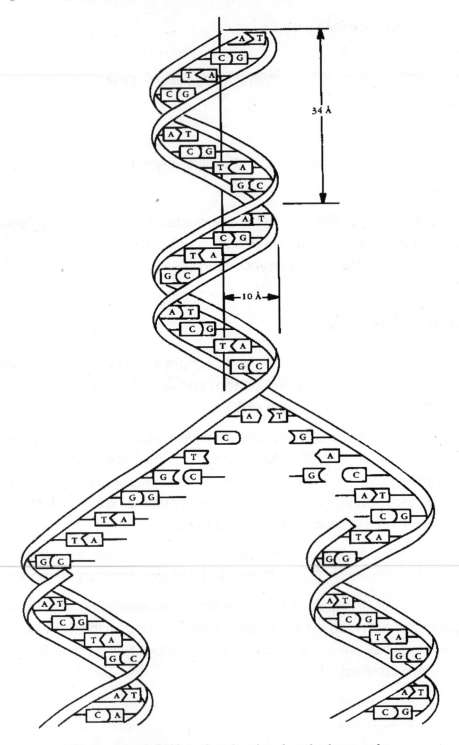

Figure 1-3. A DNA molecule, showing the base pairs.

PROTEIN SYNTHESIS

We are still left with the question of how the four nucleotides direct the synthesis of twenty amino acid to build proteins and enzymes.

Researchers found that triplet groups of nucleotides identify the different amino acids. Nucleotides can follow one another in any order. In other words, any one of the four nucleotides can be in the first, second or third position, regardless of what nucleotide came before or goes after it. This means there are 4 x 4 x 4, or 64 possible triplet combinations.

Obviously there are more triplet combinations (64) than there are amino acids (20). Well, there is redundancy built into the code, so there can be two or three different triplets that identify the same amino acid.

DNA material is held within the nucleus of the cell, while the site of protein synthesis, the ribosome, is outside the nucleus in the cytoplasm. How does DNA transmit this information outside the nucleus?

RNA molecules were found to be involved in the flow of genetic information from inside to outside the nucleus. DNA molecules produce a complementary RNA molecule, called messenger RNA (mRNA).

The process of producing the mRNA is called transcription. An enzyme, RNA polymerase, unzips a small region of DNA. It travels down the strand of DNA, using the DNA as a template to synthesize mRNA, see **Figure 1-4**. Notice in the illustration that the nucleotide Uracil replaces Thymine in the mRNA molecule. Both of these nucleotides, Uracil and Thymine, are similar chemically and both are complementary to Adenine.

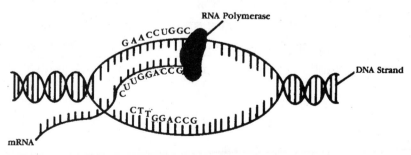

Figure 1-4. The synthesis of mRNA by RNA polymerase.

The synthesized messenger RNA (mRNA) is singled stranded and much shorter (50-1000 nucleotides) than the DNA (million nucleotides) molecule. This smaller size allows the mRNA to pass through the nucleus membrane into the cytoplasm. It is mRNA that brings the DNA information outside the nucleus for protein synthesis.

THE GENETIC CODE

The mRNA brings the information outside the nucleus for protein synthesis. We now need to know how that information is coded and synthesized. The triplet nucleotides that code specific amino acids are called codons. The codons are abbreviated using the first letter of the base in the particular nucleotide. So the codon AUG, for example, represents the bases Adenine-Uracil-Guanine. The task to find which triplet identified which amino acid became know as cracking the genetic code.

The first break in the code happened in 1961. Researchers Nirenberg and Matther found that a synthetic mRNA molecule consisting only of Uracil (making codons UUU-UUU-UUU-UUU...) produces a polypeptide consisting of the amino acid phenylalanine.

Soon all codon combinations were produced creating a genetic dictionary.

Most proteins begin with the amino acid Methionine, its codon being AUG. This codon also represents the start signal for protein synthesis. The three codons UAA, UAG and UGA do not represent any amino acid; rather, they are stop signals that terminate a protein synthesis.

Of all the amino acids, just two, Methionine and Tryptophan, are specified by a single codon. All other amino acids are specified by two or more codons. For example the amino acid Glycine is identified by any codon that begins with GG regardless of the nucleotide in the third position. **Figure 1-5** illustrates the entire genetic code.

Second base

First base		U	C	A	G	Third base
	U	UUU ⎫ PHE UUC ⎭ UUA ⎫ LEU UUG ⎭	UCU ⎫ UCC ⎪ SER UCA ⎪ UCG ⎭	UAU ⎫ TYR UAC ⎭ UAA ⎫ STOP UAG ⎭	UGU ⎫ CYS UGL ⎭ UGA STOP UGG TRP	U C A G
	C	CUU ⎫ CUC ⎪ LEU CUA ⎪ CUG ⎭	CCU ⎫ CCC ⎪ PRO CCA ⎪ CCG ⎭	CAU ⎫ AIS CAC ⎭ CAA ⎫ GLN CAG ⎭	CGU ⎫ CGC ⎪ ARG CGA ⎪ CGG ⎭	U C A G
	A	AUU ⎫ ILE AUC ⎪ AVA ⎭ AVG MET*	ACU ⎫ ACC ⎪ THR ACA ⎪ ACG ⎭	AAU ⎫ ASN AAC ⎭ AAA ⎫ LYS AAG ⎭	AGU ⎫ SER AGC ⎭ AGA ⎫ ARG AGG ⎭	U C A G
	G	GUU ⎫ GUC ⎪ VAL GUA ⎪ GUG ⎭	GCU ⎫ GCC ⎪ AUA GCA ⎪ GCG ⎭	GAU ⎫ ASP GAC ⎭ GAA ⎫ GLU GAG ⎭	GGU ⎫ GGC ⎪ GLY GGA ⎪ GGG ⎭	U C A G

Figure 1-5. The genetic code.   *Start signal for protein synthesis

TRANSFER RNA (TRNA)

Transfer RNA (tRNA) molecules perform the actual translation. The tRNA has three unpaired bases called the anti-codon on one end of the molecule. These bases are the complementary to the codon bases. When the anti-codon on the tRNA matches the codon on the mRNA, it binds with the codon on the mRNA strand. On the opposite end of the tRNA molecule is a site that can attach a single amino acid specified by the codon, see **Figure 1-6**. For each tRNA there is a specific amino acid that can bind to its site.

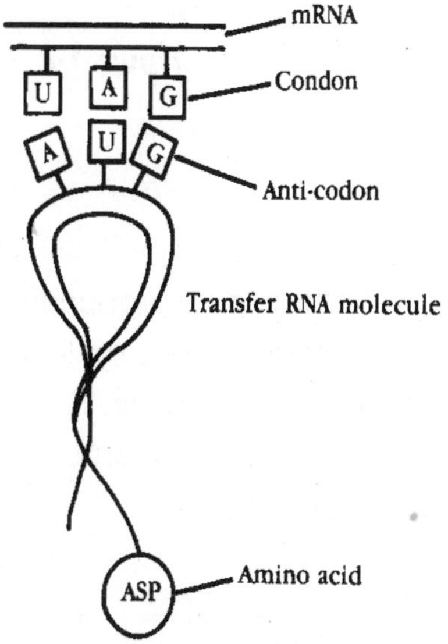

Figure 1-6. tRNA and mRNA in protein synthesis.

RIBOSOMAL RNA (RRNA)

The ribosome (ribosomal RNA) travels down the mRNA strand, linking the amino acids attached to the tRNA. As each amino acid is linked, the empty tRNA is released, see **Figure 1-7**. The released tRNA is able to pick up an amino acid and again bind to a site on the mRNA strand.

The ribosome begins protein synthesis by attaching itself to the codon AUG, which, as stated before, is the codon for Methionine and also the start message for protein synthesis. It continues linking amino acids together until it reaches one of the stop codons UAA, UAG or UGA which terminates synthesis. The completed protein is release to perform its function.

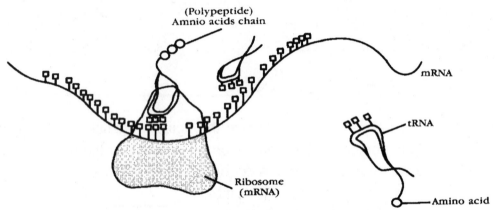

Figure 1-7. Overview of rRNA protein synthesis.

CLONES

Bacteria that are reproduced through asexual reproduction are clones. The DNA from the parent cell is replicated, with identical genetic information that is passed on to the daughter cells. Successive generations of cells in turn divide and quickly give rise to a population of genetically identical clones. All cells are derived from a single ancestral cell.

The idea of cloning a human from the DNA in one of its cells is possible, and makes for a good science fiction plot, but it is still a number of years away from being practical.

E. coli is a favorite bacteria used in genetic engineering. It is part of the normal bacterial fauna that inhabits the human colon. E. coli replicates itself once every 22 minutes. In just 11 hours it will have gone through 30 generations and created more than one billion cells.

GENETIC MANIPULATION BY VIRUSES

Viruses can attack cells and use either the cells' machinery to produce new viruses or incorporate its own genetic material into the host cells' DNA, see **Figure 1-8**. This is why it is so difficult to cure viral infections (i.e. AIDS).

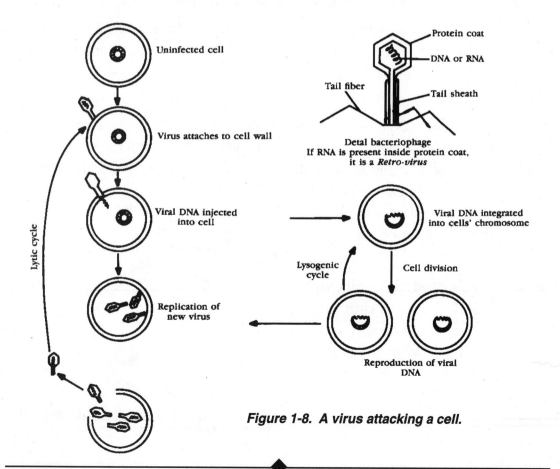

Figure 1-8. A virus attacking a cell.

In genetic engineering we use bacteria that reproduce asexually to manufacture proteins. Genetic engineering is accomplished by similar means of those used by viruses. We insert a gene of interest into the cell's genetic information. As the host cell reproduces, so does the gene of interest. When the cell goes about its normal life synthesizing proteins it needs to survive, it is also producing the proteins from the inserted genes. These proteins could be insulin, growth hormones, interferon or any genetic information we can splice into the cell DNA successfully.

FIRST STEP TOWARD GENETIC ENGINEERING

In the 1950s it was found that a certain strain of Escherichia coli (E. coli) could protect themselves from foreign introduced DNA. These cells have a primitive immune system, see **Figure 1-9**. The E. coli possesses an enzyme system that selectively destroys foreign DNA, but leaves its own DNA intact.

This protective enzyme was extracted and analyzed. This enzyme was the first "restriction endonuclease" discovered. The enzyme worked by cutting the foreign DNA into pieces. In essence, restriction endonuclease enzymes are molecular scalpels that can cut DNA.

In 1970, researchers Smith and Wilcox isolated another restriction endonuclease enzyme from Haemophilus influenza. The enzyme was named HindII. The difference with this new enzyme is that it cut DNA at predictable points, within a recognition sequence of nucleotides. This is in contrast to the other enzymes that cut the DNA pretty much at random points.

Daniel Nathans used the HindII enzyme to cut the DNA of a small virus that infects monkeys, called simian virus 40 or SV40 for short. Doing so he created a restriction map of the virus that showed where the enzyme cut the DNA.

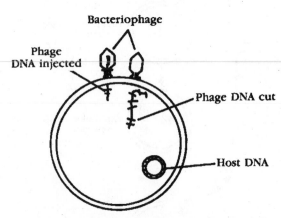

Figure 1-9. A close-up of a primitive cellular immune system.

RECOMBINANT DNA

In 1972, Paul Berg worked out a method of joining DNA molecules. Using the restriction enzyme EcoRI he cut the circular SV40 DNA virus and a small circular DNA molecule from E. coli. The restriction enzyme cut each molecule at a single point, opening the circular DNA to form strands, see **Figure 1-10**.

The DNA molecule from E. coli is particularly important. The DNA molecule used exists apart from the bacteria's main chromosomes. The small DNA molecule, called a plasmid, also has the ability to replicate and produce proteins on its own. The plasmid vector will be explained in greater detail later on.

To join the two strands of DNA together he made the ends of the DNA strand sticky; they are called, appropriately, "Sticky Ends". He accomplished this by adding a tail of 50 to 100 Adenine nucleotides to the SV40 virus using the enzyme terminal transferase. Next a tail of Thymine was added to the E. coli DNA using the same method.

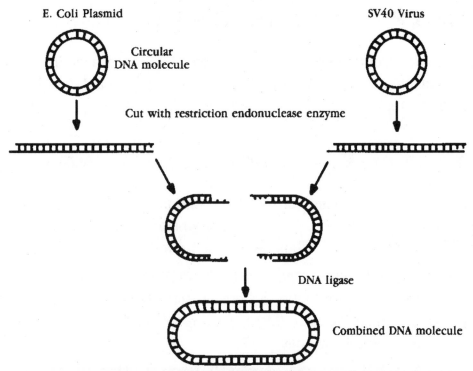

Figure 1-10. The procedure for combining SV40 virus with a plasmid.

When the two DNA molecules were mixed together, the complementary Adenine and Thymine tails paired to form one circular, recombinant DNA molecule. Two other enzymes were employed to finish the job. DNA polymerase filled any single stranded gaps and DNA ligase sealed the junction points, see **Figure 1-11**.

Later it was found out that EcoRI produced its own sticky ends, eliminating the need to add tails to the DNA strands.

Paul Berg's experiment illustrated that a restriction enzyme could be used to cut DNA in a predictable and controlled manner and that the DNA fragments from different organisms could be joined together.

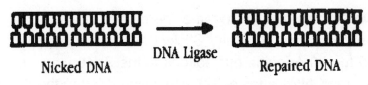

Nicked DNA   DNA Ligase   Repaired DNA

Figure 1-11. How ligase repairs breaks in DNA.

PUTTING RECOMBINANT DNA TO WORK

In 1973, Stanley Cohen and Annie Chang brought Berg's experiment further. They inserted the recombinant plasmid back into the E. coli bacteria where it was maintained and replicated itself along with the cell.

RESTRICTION ENDONUCLEASE ENZYMES

There are three major classes of restriction endonuclease. Types I and III cut DNA at sites a distance away from their recognition sequences. ATP must be provided to provide energy for these enzymes to work.

The enzyme used most often is the Type II. These enzymes cut DNA in a predictable manner on or adjacent to the recognition site. The energy required is supplied by a simple magnesium ion (Mg ++). Currently there are over 1200 Type II restriction endonuclease enzymes.

The names of these enzymes are coded as follows:

1) The first letter is usually the initial letter of the organism from where the enzyme is collected.

2) Second and third letters are usually the organism's species name.

3) Forth letter, if any, indicates the strain.

4) Roman numeral usually indicating the order of discovery.

So EcoRI breaks downs as follows:

E = genus Escherichia

co = species coli

R = strain RY13

I = first endonuclease isolated

HindIII breaks down to:

H   = genus Haemophilus

in = species influenza

d   = strain RD

III = third endonuclease isolated

MAKING E. COLI COMPETENT

Competent is a term used to describe cells brought into the proper physiological state, so that they can absorb molecules of foreign DNA from the environment. In 1970, Mandel and Higg found that E. coli becomes competent when the cells are suspended in a cold calcium chloride solution and subjected to a brief heat shock at 42°C, see **Figure 1-12**.

The precise mechanism of DNA uptake by competent E. coli cells is at the moment unknown.

In any case, this is the method that we shall use for our experiment.

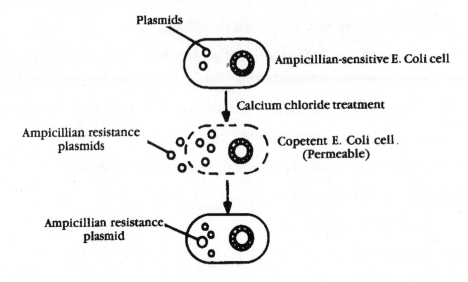

Figure 1-12. The procedure for making E. coli competent for transformation.

PLASMID VECTOR

In standard medical terminology, a vector is an organism that carries a disease from one host organism to another. In genetic engineering a vector is a DNA molecule that carries an inserted gene into a host cell.

The plasmid vector we shall use is from E. coli. Plasmid vectors range in size from 1000 to 200,000 base pairs (bp). It is a circular or looped DNA molecule. The advantage to using a plasmid DNA is that it exists separately from E. coli's main chromosome. In addition, it has the ability to replicate itself within the cell and be propagated through successive bacterial generations. The plasmid structure within the host cell carries out protein synthesis. And as in our case it will also synthesize the protein from the gene (DNA) instructions we inserted into it.

The making of a pAMP plasmid vector is illustrated in **Figure 1-13**. Although the regents are available for splicing our own plasmids and inserting a gene, we may also purchase a kit with plasmid vectors and antibiotic genes ready to be joined together and inserted into a bacterial cell.

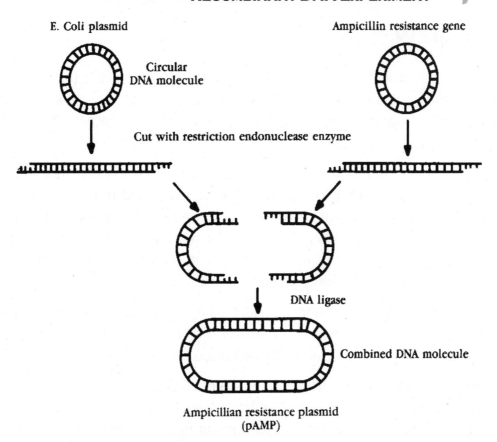

Figure 1-13. The procedure for combining the ampicillin gene with a plasmid.

GENE SPLICING EXPERIMENT

A word of caution is in order before we start. Although the materials in the kit are safe, it is important for you to follow simple procedures to keep the experiment controlled and nonthreatening.

We are working with E. coli bacteria, a favorite of gene splicers everywhere. It is not a pathogenic bacteria. In fact, it is part of the normal bacteria fauna that exists in your colon. It is rarely associated with any disease.

OVERVIEW

First we will construct three types of vector plasmids. One plasmid will contain an antibiotic resistance gene for ampicillin called pAMP. The second will contain a gene for the resistance of the antibiotic kanamycin called pKAN. The third will contain the genes of both called pAMP/pKAN.

The plasmid fractions are mixed with DNA ligase to form stable recombinant DNA. Next we use the calcium chloride procedure to inject and transform the E. coli bacteria with the recombinant DNA. To test our bacteria for uptake of the recombinant DNA we culture the bacteria on specific agar plates. One plate has ampicillin, another has kanamycin, the third has both ampicillin and kanamycin and the fourth is plain nutrient agar.

Only the bacteria that have been transformed can grow on the antibiotic treated agar. Our control E. coli group that remains untransformed will not grow on the antibiotically treated plates. But our transformed bacteria will.

As stated, you can purchase all the materials necessary separately to do this and other DNA experiments or purchase a kit. I strongly advise to purchase the E-Z Gene Splicer DNA Recombination and Transformation Kit ($64.00) from Images Company, see suppliers index in Appendix A.

The kit includes:

Vial plasmid pAMP	12 needle point pipets
Vial plasmid pKAN	18 1-ml sterile transfer pipets
3 vials ligase/ligation	4 15-ml sterile culture tubes
Culture E. coli	5 sterile inoculating loops
Vial calcium chloride	Glass cell spreader
3 agar plates	Manual
2 AMP agar plates	2 AMP/KAN agar plates
2 KAN agar plates	

The kit supplies all the tubes, pipets, plates, cultures and regents needed for this experiment.

When you receive the kit in the mail, refrigerate the culture plates (upside down) and the vials of plasmid pAMP, pKAN and calcium chloride. Freeze the vial of ligase/ligate buffer/ATP. The other materials can be stored at room temperature.

In addition to the materials provided in the kit you also need: 70-95% ethanol alcohol, small quantity of distilled water, marker for labeling culture plates and tubes, beaker or dish, 98.6°F incubator, 106°F water bath, crushed ice, small quantity of household bleach, alcohol lamp or Bunsen burner.

98.6°F INCUBATOR

Since E. coli inhabits the human gut it is not surprising that the optimum temperature for its growth is human body temperature, 98.6°F. You can use any enclosed space for an incubator. If you don't have such an area a 20 gal. glass aquarium will suffice nicely. Place the aquarium on its side with the open end facing you. Tape a piece of plastic to the top of the aquarium so that the plastic drapes down covering the open end, see **Figure 1-14**. Secure the plastic to just the top of the aquarium. In order to work inside you will need to lift the plastic up out of the way.

To heat the incubator to 98.6°F, use a standard incandescent lamp enclosed in a can or small pail. I needed a 75 watt bulb to warm the incubator to 92°F. This is lower than the optimum temperature recommended but it works fine. Start out using a low wattage 40 watt bulb and measure the temperature after 12-24 hours. Increase the wattage if necessary. If incubator becomes hotter than 98.6 F, reduce the wattage of the bulb. If it is impossible to adjust the temperature by just changing the wattage of the bulb, insert a light dimmer control to the bulb. Use the light dimmer to adjust the power to the lamp and consequently the temperature. It is better to keep the temperature a little lower than 98.6°F than above.

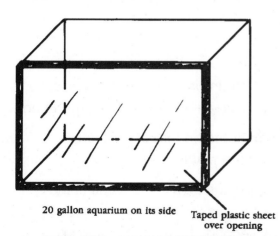

20 gallon aquarium on its side   Taped plastic sheet over opening

Figure 1-14. A simple incubator.

It's a good idea to set up your incubator before the kit arrives; this way you can start on it right away.

107.6°F WATER BATH

You only need the water bath once for 90 seconds to heat shock the bacteria. You can use an aquarium heater to bring the water to this temperature. In a pinch just run tap water, adjusting the temperature with a thermometer to 107 degrees. Adjust and keep the water running into a small container. Use water in the container to heat shock the bacteria.

SIMPLE PROCEDURES

Follow these procedures:

Wash your hands with an antibacterial soap before and after working with the materials.

Keep work space and incubator spotless. Wipe area with 10% bleach solution or a disinfectant like Lysol.

Disinfect all materials after use, such as tubes, pipets and transfer loops, by placing in 10% bleach solution.

STARTER PLATE

Our first step is to incubate and grow E. coli strain bacteria on an agar culture plate. We will need the colonies grown on this plate for the rest of the experiment.

1) Take one plain LB agar plate and mark the bottom of the plate "E. coli". Indicate the date you streaked the plate.

2) Using the wire inoculating loop, sterilize the loop in the flame of the alcohol lamp or Bunsen burner. Allow wire to get red hot, remove from flame. Hold the loop for a few seconds for it to cool, and do not place inoculating loop down, as this would contaminate it.

3) Hold vial of E. coli culture in opposite hand and remove cap. With cap removed pass mouth of vial through flame to sterilize it.

4) Push inoculating loop into the side of agar to cool loop. Then drag the loop a few times across the area of E. coli culture. Remove the loop, pass the mouth of the vial through the flame again and recap it.

5) Lift top of agar plate, marked "E. coli" just enough to perform streaking, see **Figure 1-15**.

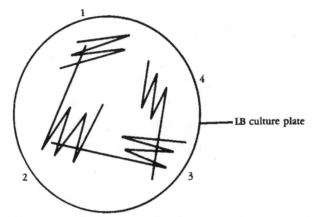

Figure 1-15. How to make streaks on a starter plate.

a) Drag loop back and forth across the agar surface. Do not stab loop into the agar. Make a few streaks across the top as in **Figure 1-15**. Replace lid between streaks.

b) Turn agar plate 1/4 turn. Reflame loop, cool loop by stabbing it into the agar plate away from first steak. Pass loop once through first streak and continue in zig-zag pattern.

c) Turn agar plate 1/4 turn. Reflame loop and proceed as before, except draw loop once through secondary streak.

d) Turn agar plate 1/4 turn. Reflame loop and proceed as before, except draw loop once through third streak, and make a final zig-zap pattern.

e) Replace lid on culture plate.

6) Reflame loop before putting it down. This prevents contaminating work space. Make this a habit.

7) Place plate upside down in incubator. This prevents condensation that may collect on the lid from falling into the agar and smearing the E. coli colonies. Incubate plate for 12-24 hours.

8) After initial incubation, remove from incubator and allow colonies to grow 1-2 days at room temperature.

Do not over-incubate the culture or the E. coli will over-grow on the plate.

LIGATION OF DNA

In this procedure we are linking the fragmented DNA molecules together. The regents pAMP and pKAN have both the plasmid DNA strands from E. coli bacteria, plus the antibiotic DNA fragment. The ligase links the two DNA molecules via the formation of a phosphodiester bond. See simple structure of DNA molecule under nucleotides.

It is interesting to note that this procedure forms many different types of hybrid molecules, such as plasmids composed of more than two fragments. However, only those that form properly will be maintained and expressed in the cell.

The ATP in the ligation solution provides the energy for the reaction joining the nucleotides together.

1) You will need the three vials to contain the 20 μl of ligation buffer ATP/ligase. Label one tube +pAMP/KAN, label another +pAMP and the last tube +pKAN.

2) To measure quantities use a sterile needle-nose pipet, see **Figure 1-16**. Use a fresh needle-nose for each regent.

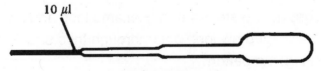

10 μl

Figure 1-16. A detail of the needle-nose pipet.

   a) To the +pAMP/KAN tube, add 10 μl of pAMP and 10 μl of pKAN.

   b) To the +pAMP tube, add 10 μl of pAMP and 10 μl of distilled
      water.

   c) To the +pKAN tube, add 10 μl of pKAN and 10 μl of distilled
      water.

3) Close tube tops and tap tube bottoms on table to mix regents.

4) Incubate tubes at room temperature for 2-24 hours, see **Figure
   1-17**.

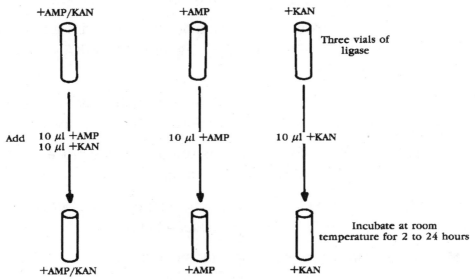

Figure 1-17. DNA ligase procedure.

TRANSFORM E. COLI WITH RECOMBINANT DNA

In this procedure we are making the E. coli cells competent which allows the cells
to uptake the recombinant DNA we made in the preceding procedure.

1) Get four sterile 15 ml tubes. Label one "+pAMP/KAN", label the
   second "-pAMP/KAN, the third "+pAMP" and the forth "+pKAN".
   Using a sterile transfer pipet, see **Figure 1-18**, add 250 μl of cold
   calcium chloride to each tube. Place four tubes into beaker or dish
   with crushed ice.

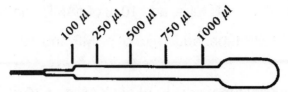

Figure 1-18. A detail of the transfer pipet.

2) Using a sterile plastic inoculating loop, transfer one or two colonies of E. coli from the starter plate to the +pAMP/KAN tube. Be careful not to transfer any agar from the plate along with the cell colonies.

    a) Immerse the loop into the calcium chloride solution. Tap against the side of the tube to dislodge the cell mass.

    b) Suspend cells in the solution by repeatedly pipeting in and out with a sterile transfer pipet. Return +pAMP/KAN tube to ice.

3) Transfer cell colony to -pAMP/KAN tube using the same procedure as describe in step 2. Return -pAMP/KAN tube to the ice.

4) Transfer cell colony to +pAMP tube using the same procedure as described in step 2. Return +pAMP tube to the ice.

5) Transfer cell colony to +pKAN tube using the same procedure as described in step 2. Return +pKAN tube to the ice.

6) Using a fresh needle-nose pipet for each transfer:

    a) Transfer 10 µl of ligated +pAMP/KAN to +pAMP/KAN culture tube.

    b) Transfer 10 µl of ligated +pAMP to +pAMP culture tube.

    c) Transfer 10 µl of ligated +pKAN to +pKAN culture tube.

    d) Do Not transfer any material into -pAMP/KAN culture tube.

7) Place all tubes back in ice and let them incubate on ice for 15 minutes.

8) Following the 15 minute ice incubation it's time to heat shock the E. coli cells to make them competent. Remove all the tubes from the

ice and immediately immerse them in the 107°F water bath for 90 seconds. Then return all the tubes directly into the ice again. Let tubes stay on ice for 3-4 minutes.

9) Using a sterile transfer pipet, add 250 µl of Luria broth (LB) to each tube. Gently tap tube with finger to mix broth and place tubes in incubator, at 98.6°F for 3-6 hours, see **Figure 1-19**.

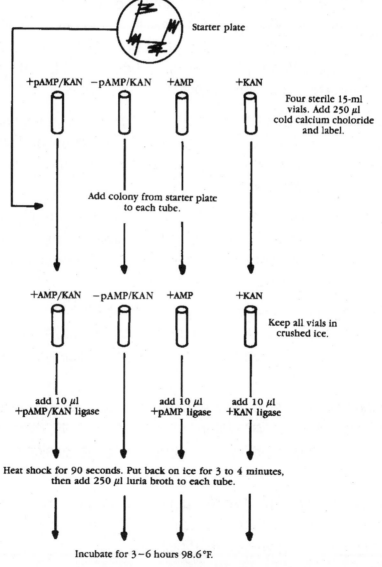

Figure 1-19. The uptake of recombinant DNA by competent E. coli.

PLATING THE RESULTS

In this procedure we will check to see if we have successfully incorporated our recombinant DNA into the E. coli bacteria.

The culture plates supplied with the kit are marked showing the agar medium.

1) Using the plates supplied in the kit label them as follows:

a) Label one LB plate "+". Label one LB plate "-". Label one LB/AMP/KAN plate "+pAMP/KAN". Label one LB/AMP/KAN plate "-pAMP/KAN". Label one LB/AMP plate "+pAMP". Label one LB/AMP plate "+pKAN". Label one LB/KAN plate "+pKAN". Label one LB/KAN plate "+pAMP".

2) Using a sterile transfer pipet add 100 µl of the cell suspension from the -pAMP/KAN culture tube on the -LB/AMP/KAN plate and an other 100 ul on the -LB plate. Spread cells over the surface of the agar using the following procedure.

a) Dip glass spreader in the ethanol alcohol, and ignite the alcohol using the Bunsen burner or alcohol lamp. After alcohol burns off the spreader it is sterile for use. Use spreader to evenly distribute cells over the agar.

3) Using another sterile transfer pipet add 100 µl of the cell suspension from the +pAMP/KAN culture tube to the +LB/AMP/KAN plate and another 100 µl on the +LB plate. Spread cell suspension as outlined in 2a.

4) Using another sterile transfer pipet add 100 µl of the cell suspension from the +pAMP culture tube to the +pAMP plate, and another 100 µl to the +pKAN plate. Spread cell suspension as outlined in 2a.

5) Using another sterile transfer pipet add 100 µl of cell suspension from the +pKAN culture tube to the +pAMP plate and another 100 µl to the +pKAN plate. Spread cell suspension as outlined in 2a.

Allow plates to set up for 10 minutes, then wrap together with tape. Place plates upside down in 98.6°F incubator. Incubate plates for 12-24 hours.

RESULTS

Figure 1-20 illustrates the results of the experiment. If the E. coli growth is too dense to count, record lawn.

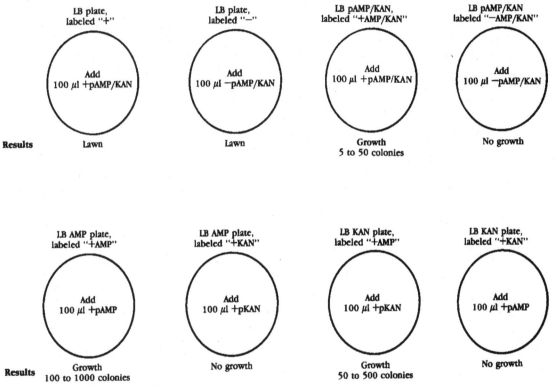

Figure 1-20. The results of transformed E. coli on culture plates.

1) The LB plates labeled "+" and "-" show that both the transformed E. coli and natural E. coli are viable and grow equally well.

2) The LB plates that contain ampicillin and kanamycin labeled "+AMP/ KAN" and "-AMP/KAN" show that only the E. coli that have been transformed with both antibiotic resistant genes can grow. The natural E. coli fails to grow in this medium.

3) The LB plates that contain the ampicillin illustrate that the gene for pKAN does not contain a resistant gene for ampicillin.

4) The LB plates that contain kanamycin illustrate that the gene for pAMP does not contain a resistant gene for kanamycin.

By measuring the growth of the colonies we can determine that ligation of two genes pAMP and pKAN is more rigorous (5-50 colonies) than either single ligation. Further, that ligation of the pKAN gene is more rigorous (50-500 colonies) than ligation of pAMP gene (100-1000 colonies).

E-Z Gene Splicer DNA Recombination and Transformation Kit is available from Images Company. See Appendix A.

CHAPTER 2

◆

GENETIC EVOLUTION

In this experiment we will trace the evolution of simple yeast cells through genetic mutations. Yeast are unicellular fungi that reproduce asexually through budding. The mutations are a manipulated "natural selection" caused by placing the yeast in a hostile environment where it must mutate in order to survive.

As with the recombinant DNA experiment a kit is available, and I advise to purchase the Yeast Evolution Kit from Hermant Chikarmane rather than buy the materials separately, see suppliers index in Appendix A.

The kit includes the following materials:
    5 Culture Plates Labeled:

Master	Vial of Yeast Culture
Adenine Negative A	4 Sterile Spreaders
Adenine Negative B	4 Sterile Pipets
Copper A	40 Sterile Toothpicks
Copper B	

When you receive the kit refrigerate it until you use it.

In addition to the materials in the kit you will need a marker, paper, 10% solution bleach, disinfectant such as Lysol, and antibacterial soap.

PROCEDURES

To keep the experiment accurate and safe follow these simple procedures.
1) Wash you hands before and after you work with the kit using an antibacterial soap.
2) Keep work area spotless. Clean area with disinfectant before you begin experiment.
3) Discard all used materials in 10% bleach solution.

The yeast strain we will be using is Saccharomyces cerevisiae. This yeast strain has a mutation in its ade1 gene. This prevents the synthesis of an enzyme the yeast requires to make Adenine, which if you remember from the recombinant DNA experiment is one of the base molecules for DNA. Without this ability to manufacture its own Adenine the yeast cannot replicate unless Adenine is provided.

EXPERIMENT 1

Figure 2-1 illustrates the entire experiment.

The Adenine Negative plates do not contain any Adenine in the nutrient agar. The Copper plates contain copper salts which are poisonous to yeast.

In the first part of the experiment we inoculate one Adenine Negative plate and Copper plate with yeast. Only those yeast that mutate can replicate and form colonies on these two "Hostile Environment" plates. These mutations have been "naturally selected" by their environment.

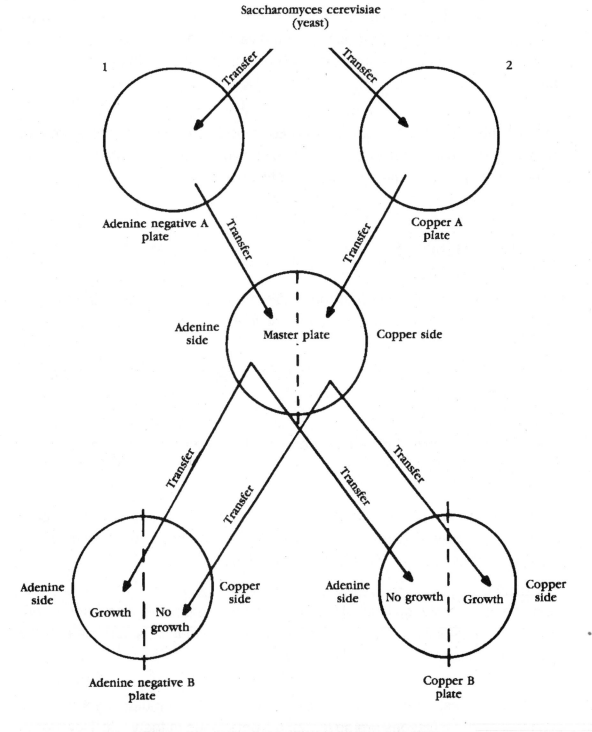

Figure 2-1. An overview of the culture experiment.

Shake up the vial that contains the yeast cells. Using a sterile pipet, put three drops of yeast culture on the Adenine Negative A plate, and three drops on the Copper A plate. Spread the drops carefully over the surface of the agar using the spreader.

Incubate the plates for 2-5 days. When colonies appear on the plates we can proceed to the second part of the experiment. If you wish to hold off and stop the experiment for a short time refrigerate the two plates.

EXPERIMENT 2

In this section we will show that the mutation is transmitted to their offspring, regardless of the fact that the mutation is no longer advantageous or required.

The master culture plate has all the nutrients yeast requires. No poisons in the form of copper salts are added to the agar.

Divide the master culture plate in two equal halves using your marker. Label one side Adenine and the other Copper.

Using a sterile tooth pick, lift a colony off the Adenine Negative plate from experiment one, and streak it across the Adenine side of the master culture plate. Make a few more streaks on this side of the plate.

Using another sterile toothpick, lift a colony off the Copper A plate from experiment one, and streak it across the Copper side of the master culture plate. Make a few more streaks on this side of the plate.

Incubate the master plate for a few days. Notice that the mutated yeast grows even though the mutations are no longer advantageous to the yeast. How do we know the yeast are still in their mutated form? The answer lies in experiment 3.

EXPERIMENT 3

In this section we shall prove the yeast are still in their mutated form, further that the naturally selected mutations are independent of one another.

Divide the Adenine Negative B and Copper B plates in half using a marker. Label one side Adenine and the other Copper.

Look at **Figure 2-1**. Using a sterile tooth pick, transfer a colony from the Adenine side of the master plate and streak the Adenine side of the Adenine B plate and the Copper B plate. (Use a colony for each plate.)

Using a sterile toothpick transfer a colony from the copper side of the master plate and streak the copper side of the Adenine B plate and the Copper B plate. (Use a colony for each plate.)

Incubate the plates for a few days.

RESULTS

The Adenine mutated yeast from the master plate grows on the Adenine side of the Adenine B plate but not on the Copper plate. This shows that they retain their mutated trait.

The copper mutated yeast from the master plate grows on the copper side of the Copper B plate but not on the Adenine plate. This shows that they also retained their mutated trait.

EVOLUTION AND SEX

Biologists suggest that sex evolved because it creates a genetic diversity. The random shuffling of the genes can bring forward favorable traits, that through "natural selection" promotes the species.

Well, as we have seen, simple organisms seem to mutate and adapt to their environment quite well, without mixing genes though sex.

Evolutionary biologist Richard Michod of the University of Arizona has developed a different possible reason for sex. He believes sex evolved for the species to repair their damaged DNA. The DNA gets damaged naturally through the years, from the environment, through the food, chemicals, UV from the sun, etc.

It would be highly unlikely for two members of the same species to have their DNA molecules damaged at the same points. So sex got started as a recombination of the two parents' DNA to repair any potentially damaged to the offsprings' DNA.

Does it wash? Yes. In one experiment two specific mutated E. coli colonies were grown. One colonie's genes were damaged so it could not synthesize a particular amino acid, let's call this amino acid A. The other colony had a different gene damaged where it couldn't synthesize a different amino acid, let's call this amino acid B.

When both colonies were placed on a culture plate whose nutient media lacked both amino acids A & B, some colonies formed that could exist on the culture plate.

Apparently the E. coli culture were able to recombinate its genes with one another and repair the damaged genes. The process is called conjugation, and is analogous to sexual mating in higher organisms.

CHAPTER 3

---◆---

NEGATIVE ION GENERATOR

Air is the most important necessity to our survival. Think about it. A person may survive a few days without water, a little longer without food, but deprived of air to breathe survival time is measured in minutes.

The quality of air, or lack of it, surrounding many cities has become such an important consideration that many local news stations provide an air quality report along with the weather forecast. Air pollution is so commonplace now that new words are created to describe it. The word smog for example, is a contraction of the words smoke and fog.

Today there are questions involving global air quality: the increasing CO_2 level (0.03%) and the "Green House Effect", the Ozone layer depletion and increased UV, Acid rain, etc.

ION EFFECT

Long before there was any talk or concern about air pollution and such, people have had a reaction to the quality of air in regard to its ionization. This idea was popularized by Fred Soyka who in the 1970's wrote a book titled *The Ion Effect*. Mr. Soyka studied natural occurrences of negative and positive ionized air. His findings and inquiries demonstrated that negative ionization had substantial health benefits.

To summarize a few points in this book, negative ions help elevate mood, enhance physical performance and training, and sterilize harmful airborne bacteria. An abundance of positive ions on the other hand can be held responsible for a number of low grade medical problems, such as fatigue, headaches and anxiety.

Air is composed principally of nitrogen (78%) and oxygen (21%). In addition, air is typically full of positive and negative ions (app 5:4 ratio). When the balance of ions fall heavily into either region the effects of the air ionization become apparent in biological systems.

ADDITIONAL RESEARCH

The first question that entered my mind when I began this project was whether or not those early reports are still considered accurate. I answered this question for myself by surveying approximately one hundred scientific reports on the effects of negative ions gathered worldwide from 1973 through the present. I can report that out of my survey approximately 80% of the citings support beneficial effects of negative ions. Of the majority balance of reports, greater than 19% described no effect, and a few, less than 1%, detailed some detrimental effect. Since the preponderance of the evidence still supports positive effects of negative ions I felt this to be a worthwhile project. Before we begin our actual project, let's take a look at some of these later reports.

◆

POSITIVE EFFECTS

Learning enhancement in normal and learning disabled children. The task used to test the children was a dichotic listening test.[1]

Negative ions were used to decrease amounts of Radon in building atmosphere.[2]

In one animal study 1279 calves were broken into two groups, one 649 head and the other 630 head. Negative air ionization was used to test for a prophylactic effectiveness against respiratory diseases. The results were remarkable: In the treated group (649 head) 45 calves became sick and 3 died. In the control group (630 head) 621 became sick and 33 died.[3]

Reduction (40-50%) of microbial air pollution in dental clinics.[4]

Test using college students showed improved performance on a visual vigilance task.[5]

This is by no means exhaustive, it's just a sampling of the scientific literature available. But if this is the case it would be to our benefit to improve the quality of air that we breathe with a negative ion generator.

ASTHMA

The reports on the effects of negative ions on asthma vary. Some describe beneficial effects, some state no effect and one claims a negative effect. All the experiments and consequential reports have been performed by reputable scientists. Clearly there is still a need for additional research before a determination can be arrived at definitively.

SPORTS

Recently an article in a popular body-building magazine put forward a theory that negative ions may be used to improve physical performance.[6] It stated further, that Soviet athletes may already be using negative ions which could explain their superior recuperative powers in athletic meets worldwide. (This idea was also reported in Fred Soyka's book.)

There is at least one report that I found that does indeed support this hypothesis. In 1983 it was reported that chickens raised in a negative ionized atmosphere showed improved anabolic processes. The chickens raised in negative ionized air had an overall greater weight than the control group. This in spite of the fact that the quality and quantity of the feed were the same for both groups. The meat of the treated group had a higher protein and essential amino acid content. In addition, higher concentrations of vitamins E and A were found in the liver.[7]

NO HEALTH BENEFITS CLAIMED

All commercial negative air ionizers sold in the United States are sold as air purifiers only. Despite the numerous scientific reports reporting positive health benefits, no manufacturer of negative ion generators can make any health benefit claims without running afoul with the FDA. (Neither will I.) That is why the documentation supporting this article is listed in the Bibliography, for you to verify the research on your own if you wish and to make your own decision.

SUSCEPTIBILITY

Some individuals are more susceptible to the effects of air ionization than others. Regardless of whether you feel any of the reported effects, if the scientific literature is to be believed it will in the least help purify the air.

ION GENERATOR

The negative ion generator is fairly straightforward. The main circuit is a high voltage AC device. The high voltage lead of the device is connected to a 10 kV high voltage diode, which permits negative voltage to pass to a small capacitor bank. This negative high voltage is then connected to conductive object that has or comes to a sharp point. The sharp point enhances the negative discharge into the surrounding atmosphere. In addition, a small electric fan is incorporated to provide an airflow past the discharge point and into the surrounding air.

The circuit (see **Figure 3-1**) uses a standard 555 timer to generate square wave pulses. The pulses are applied to the base of the TIP 120 NPN Darlington transistor. The Darlington provides sufficient current to the base of the 3055 power transistor to power the high voltage autotransformer. The high voltage lead off the transformer is connected to a 10 kV high voltage diode. Notice the polarity of the diode. It is biased to allow negative voltage to pass to the high voltage capacitors and discharge needle. For the sharp pointed object you can used a sewing needle. An alternate to using a needle for a discharge point is a small piece of 22 Ga. stranded wire. Strip off about 1/2 inch from one end and separate the fine copper

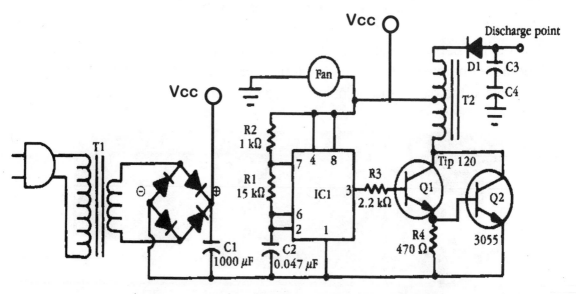

Figure 3-1. A schematic of a negative ion generator.

wires so that they are evenly dispersed (see **Figure 3-2**). Each wire end should now behave as a discharge point. The other end of the wire connects to the negative high voltage lead.

You can use any enclosure large enough to hold all the components. I'd recommend using a plastic enclosure if one is available.

The fan is situated in the enclosure to pull air in past the discharge point(s) and out through the opening hole at the top of the enclosure, see **Figure 3-2**. I put the

Figure 3-2. A top view of the ion generator.

HV diode, HV capacitors and discharge needle (or point(s)) on a small piece of PC board, separate from the main circuit board, see **Figure 3-3**. This made it easy to place the discharge point in a good location relative to the fan.

Figure 3-3. An inside view of the ion generator.

If you decide to put any screen or covering on the output fan hole I advise that it should be nonmetallic or plastic in nature. I think that using a metal screen would severely cut the efficiency of the generator, because as the negative ions came into contact with the metal screen they would be neutralized.

Remember to place a few air holes in the side or bottom of your enclosure for the fan to draw air in.

When testing the circuit if you see any arcing or discharge from the HV transformer or HV capacitors, spray the area with a little "No Arc" spray available from Radio Shack. Allow the material to dry before testing the unit again.

RESEARCH PROJECTS

Here are a couple of topics where you can do additional research with negative ions.

PLANT GROWTH STIMULATION

An interesting experiment to perform is to measure the effect of negative ions on the growth of plants. It has been reported that negative as well as positive ions increase the growth rate of plants.

IMPROVED GAS MILEAGE

It has been reported that negative ions improve the efficency of internal combustion engines and at the same time reduce emmissions. It accomplishes this by improving the burning of gasoline in the cylinders.

SWIMMING POOLS

By aerating swimming pool water with negative atmospheric ions the use of chlorine can be significantly reduced.

PARTS LIST

◆

	ITEM		SOURCE
TR1	120 VAC/12V 1.2A Step-down Transformer	# 273-1352	Radio Shack
BR1	4A Bridge Rectifier	# 276-1171	Radio Shack
C1	1000 μF	# 272-1032	Radio Shack
C2	.047 μF	# 272-1068	Radio Shack
IC1	555 Timer	# 276-1723	Radio Shack
R1	15 Kohm 1/4 Watt	# 271-1337	Radio Shack
R2	1 Kohm 1/4 Watt	# 271-1321	Radio Shack
R3	2.2 Kohm 1/4 Watt	# 271-1325	Radio Shack
R4	470 ohm 1/4 Watt	# 271-1317	Radio Shack
Q1	TIP 120 NPN Darlington	# 276-2068	Radio Shack
Q2	3055 Power Transistor	# 276-2041	Radio Shack
Fan	12 VDC Fan	# 276-243	Radio Shack
TR2	High Voltage Transformer		Images Company
D1	10 KV Diode		Images Company
C3, C4	6 KV Cap		Images Company
Misc	Enclosure, line cord, switch, TO-3 socket & heat sink		

◆

CHAPTER

BIO-FEEDBACK &
LIE DETECTOR DEVICE

Bio-feedback devices allow individuals to train or control an aspect of their autonomic physiology. Autonomic, in other words, means automated and controlled by lower brain functions and therefore not under our conscious control. This belief, however, is outdated - it proved to be a fallacy. Training a person to control such physiology as EEG (brain waves), EKG (heart rate), blood pressure, and tension level proved to be possible once a method became established of showing a person (feedback) the physiological changes they are trying to control in real time.

It is interesting that the control of these functions cannot be accurately taught verbally. It is the training or the conditioning of the body using a bio-feedback device to feel a particular way or to get a feeling that controls these functions.

The types of bio-feedback devices available are numerous. The type that we will build in this chapter is a galvanic skin resistance bio-feedback device, see **Figure 4-1**. Galvanic skin resistance is a good indicator of stress level in the subject. But perhaps its most famous attribute is its use as a lie detector.

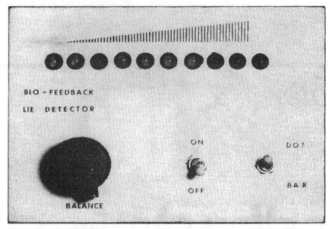

Figure 4-1. The bio-feedback unit.

CIRCUIT DESCRIPTION

The circuit (**Figure 4-2**) is broken down into two main parts. The front end consists of the op-amp and resistance bridge. This is the actual bio-feedback circuit. The back end of the circuit, the 3914 IC and ten LEDs, make up the display section.

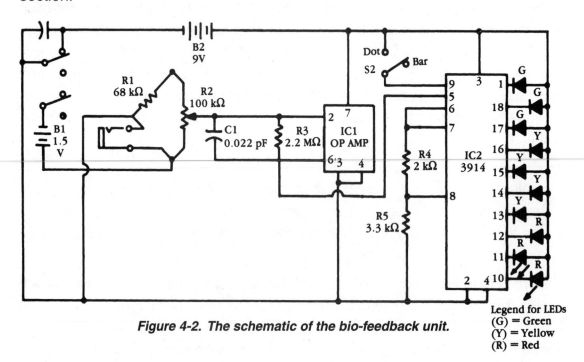

Figure 4-2. The schematic of the bio-feedback unit.

BIO-FEEDBACK SECTION

The advantages this galvanic resistance device has over previous designs is that the op-amp we are using requires a single ended +9V power supply, in contrast to the standard 741 op-amp that requires a bipolar 9V power supply. This really simplifies our circuits power requirements. We do require one 1.5V AA battery to supply power to the resistance bridge.

Circuit operation is straightforward. Looking at the schematic, examine the resistance bridge that consists of the electrodes, 1.5V battery, 68 Kohm fixed resistor and 100 Kohm potentiometer. When the electrodes are attached to the subject, the subject's resistance becomes part of the bridge. The bridges can be balanced using the 100 Kohm potentiometer. The output from the bridge is fed to the input to the op-amp. The op-amp is set up as a difference amplifier. Once balanced, this amplifier will amplify any minor change in the subject's resistance.

The output of the op-amp is directed to the input of the 3914 chip. The 3914 chip reads the voltage from the op-amp and converts it into a digital display using the ten LEDs. Two types of display are available from the 3914; bar and dot.

ELECTRODES

Silver is considered one of the best electrode materials around. We can capitalize on this by using two U.S. dimes for our electrodes. The cable we'll use is a shielded 2-conductor. The shielding is a copper braided wire that surrounds two insulated wires in the center of the cable. Remove about 2 inches of the outer cable jacket, and separate the shielding from the insulated wires. Strip 1/2 inch of insulation off the center wires.

Soldering the wires to the dimes is a little tricky if you haven't done much soldering, see **Figure 4-3**. Place the tip of your soldering iron on the coin, and keep it

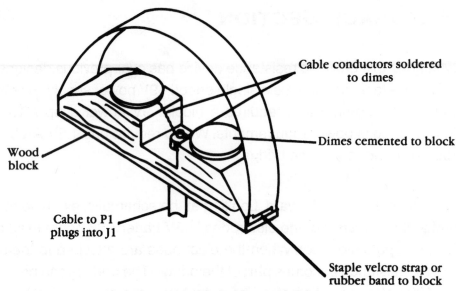

Cable conductors soldered
to dimes

Dimes cemented to block

Wood
block

Cable to P1
plugs into J1

Staple velcro strap or
rubber band to block

Figure 4-3. The electrode holder.

there till the coin becomes hot enough to melt solder on it. This takes about 1-2 minutes of continuous heating. At this point melt a small puddle of solder on the coin, and place the bare end of one of the insulated wires into the puddle. Remove the soldering iron from the coin. Keep the wire in place until the solder solidifies. Repeat the procedure for the other coin, but solder the shielding along with the wire to this coin. This will be the ground electrode.

The palm of the hand is very sensitive to galvanic changes, so it is therefore the area of choice. To secure the dime electrodes to the palm of the hand I made a small palm-fitting electrode holder out of 3/4 inch pine wood. You only need a couple of square inches practically any piece of scrap can be used, see drawing. After cutting the wood to the proper shape, drill a 1/4 inch hole through the center as shown to feed the cable through. Then glue, epoxy or hot glue the electrodes to the wood block as shown. To finish off the hand electrode, attach a rubber band or elastic material to the base so that it covers the electrodes. This materials is what will secure the holder to your hand.

CIRCUIT CONSTRUCTION

The circuit is fairly simple. If you look at the photograph of the circuit board (**Figure 4-4**) you'll see I used ribbon cable to connected the LEDs to the 3914. Ribbon cable isn't necessary but it helps keep the LEDs in proper order. The bar/dot display mode switch simply connects pin 9 of the 3914 to the +V, or lets it float.

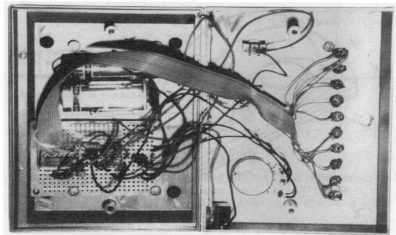

Figure 4-4. An internal view of the bio-feedback unit.

CIRCUIT OPERATION

Attach the electrodes to your subject's hand by placing the wood electrode holder in the palm and the rubber band around the hand, see **Figure 4-5**. Turn on the circuit, then adjust the balance pot so that the LED graph is lit approximately midway. You will notice that when adjusting the balance pot the LEDs jump very quickly when you reach the balance area. A soft touch is required in rotating the pot around this area.

If your subject is a little nervous you may have to adjust the balance a few times as they relax. When everything is stable, have the subject quickly inhale and exhale deeply. This should cause a rise in the LED graph that gradually returns to the previous level. If you get this result the circuit is operating properly and you're ready to go.

If you get the opposite reaction, i.e., the LED graph dips, the battery in the bridge section is reversed.

In order to test the circuit you must have some resistance connected across the electrodes or the LEDs will never light.

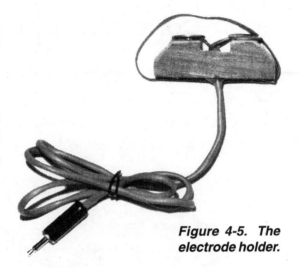

Figure 4-5. The electrode holder.

BIO-FEEDBACK

To use this device in a bio-feedback mode for relaxation and tension reduction, set the balance pot to light the graph in the upper portion. As you relax or reduce tension the body's resistance increases, which will be seen as an gradual downward sloping of the graph. When you reach the bottom you can read just the balance pot to bring it back up and then try to bring it down again.

LIE DETECTOR

To use this device as a lie detector set the graph on the lower portion of the graph. Any stress will cause the graph to rise. There is a delay between question and response of approximately 1.5 seconds. Remember, this device is for entertainment purposes only. Even full-fledged lie detectors are fallible, and it could be the nature of the question regardless of the answer that would cause a stress reaction.

PARTS LIST

	ITEM	SOURCE
IC1	CMOS OP-Amp	Images Company
R1	68 Kohm	Radio Shack
R2	100 Kohm Potentiometer	Radio Shack
R3	2.2 Meg	Radio Shack
R4, R5	2.2 Kohm	Radio Shack
C1	.047 µF	Radio Shack
SW1	DPDT	Radio Shack
Misc	1.5V Battery and Holder, 2 Dimes for use as Electrodes, Case with PC Board	

CHAPTER 5

♦

ELF
Monitor

There is a growing concern over the possible health hazards of low frequency electromagnetic emissions. When the story first broke in this country it concerned electromagnetic fields given off from overhead power line transformers.

The reasoning that followed was that for most of us unless we work in the electrical/electronic fields or live in close proximity to power lines, we could consider ourselves unaffected and relatively safe. But new evidence suggests that this really isn't the case. It appears that the ELF (extremely low frequency) magnetic fields given off by many of our household appliances and computer monitors can be of sufficient strength to be considered potentially hazardous.

For those of us who dabble in electronics, or earn our livelihood in electronics or a related field, the concern becomes more relevant. This being the case we should at least be aware of what research is taking place and what has been reported so far. And if after reading the following material you feel the concern to be legitimate, you can construct the ELF monitor to check and modify your environment.

WHY WASN'T IT SOONER

The question needs to be asked, if ELF radiation does present a health hazard why has it taken so long for anyone to uncover it? To answer this question we must look at how scientists first interpreted any potential biological hazards from low frequency magnetic fields.

To begin, it was originally believed that weak low frequency fields could not have a significant impact on living systems. This belief was based upon the amount of thermal energy the ELF fields could produce in biological tissue or cells. The energy transmitted is much smaller that the normal thermal energy generated internally by the cells' own metabolic processes. In addition, the quantum energy of the fields is far too low to break any chemical or nuclear bonds in the tissue. This being the case, they felt DNA structure to be safe from mutating. Finally, the electric field of the body is much greater than any induced field from the ELF. Looking at all these factors, it's easy to understand why scientists and the scientific community in general quickly dismissed epidemiological studies that described a statistical significant hazard associated with ELF as being flawed in one way or another.

In defense of the scientific community, which has been portrayed by the press as a bunch of hacks, or bureaucratic puppets controlled by various government agencies or industrial power companies genuflecting for grants, the reason for the quick dismissal was one of disbelief, not clandestine action for a mass cover-up. Although in truth, a few scientists have stepped over the line and maligned good researchers based on profit and loss concerns of their employers. These scientists are few in number and we should not condemn the entire scientific community based upon their isolated unethical endeavors. Most scientists by nature of being scientists, must remain open minded to new discoveries as they happen or they are quickly allotted a platform in a museum.

THE REAL DEAL

Although the mechanism by which ELF fields impact on biological tissue is not exactly known, it has been shown unequivocally that cellular tissue is affected. The best research to date shows the cell's membrane or receptor molecules in the membrane to be sensitive to extremely weak low frequency magnetic fields.

Some of the effects reported so far include changes in the flow of ionic compounds through the cellular membranes, changes in DNA synthesis and RNA transcription and the response of cells to signaling molecules such as hormones and neurotransmitters. In addition, changes have been noted in the kinetics of some cellular biochemical reactions.

Not all ELF produces all these effects; it's more complex than that. Some effects are noted at discrete frequency and amplitude of the field. Others depend upon the strength and orientation of an ambient DC magnetic field. Still others require a threshold determined more by exposure time than field strength.

It has been emphasized in the press on the inherent danger or increased likelihood of various forms of cancer with exposure to ELF. As stated previously, the quantum energy of these fields isn't sufficient to produce any type of chromosomal damage. Simply, what this means is that the ELF doesn't initiate cancer. The association to the increased incidence of cancers involves its promotion after the cancer has been triggered by another agent. The promotion of cancer is caused by the ELF suppression on the body's immune system (see cell response to ELF above). In addition at the cellular level it has been determined that the ELF fields increase the production of the enzyme ornithine decarboxylase, which has been cited to support the promotion of cancer in the body.

THE EVIDENCE

As studies progress more information shall be forthcoming. Here is a short list of reported events that indicates the potential health hazards of ELF fields.

1972 Soviet researchers link electromagnetic fields with low grade health problems such as fatigue and headaches.

1977 USA Robert Becker, physician and biophysicist Andrew Marino testified before N.Y.S. Public Service Commission about the results of their experiments which showed negative health effects due to exposure to ELF fields.

1979 USA Nancy Wertheimer and epidemiologist and physicist Ed Leeper publish a study which shows statistical link between childhood cancers and the proximity of certain types of high current power lines to the home.

1982 USA Washington State study examined the data for 438,000 deaths of workers in Washington State occurring between 1950 and 1979. The results of the study found that leukemia deaths were elevated in 10 out of 11 occupations where the workers were exposed to ELF fields.

1986 Sweden Dr. Bernard Tribukait, a professor of radiobiology at the Karolinska Institute in Stockholm reported that the fetuses of mice exposed to sawtooth shaped electromagnetic pulsed fields had a greater incident of congenital malformation than unexposed mice. The sawtooth waveform is a typical waveform generated in CRT monitors.

1988 USA Maryland Department of Heath and Hygiene found an unusually high rate of fatal brain cancer among men employed in electrical occupations.

1989 USA Johns Hopkins University found an elevated risk of all cancers among N.Y. Telephone Company cable splicers. On site reading of the ELF field showed exposure to 60 Hz ELF of approximately 4.3 milligauss.

1990 USA David Savitz, epidemiologist of the University of North Carolina has determined through a study that pregnant women who use an electric blanket have children who have a 30% increased risk of cancer as compared to children whose mother didn't use an electric blanket. See also cluster miscarriages under VDT paragraph and Nancy Wertheimer study of increased miscarriages of women who used an electric blanket under Other Precautions Around the Home.

NOT ALL THE NEWS IS BAD

So far I have concentrated on the negative effects of the 60 Hz ELF fields. But you should know that there are positive medical uses to ELF fields. Robert Becker had discovered that ELF fields when appropriately applied (specific frequency and amplitude) can promote healing and therapeutic responses in tissue (see Dr. Becker under 1977 evidence). The ELF fields appear to be a double-edged sword, being able to cure as well as kill.

COMPUTER MONITORS

Concern over televisions and computer monitors (which are closely related in operation and technology) is nothing new. A number of years ago there was a concern whether radiation given off by color televisions could have a negative impact on health. This concern was based primarily on ionizing radiation, (low level X-rays) whose intensity fell off dramatically a few inches away from the TV screen, and turned out to be incidental. But more insidious than this overt obvious threat is one that has passed unnoticed until quite recently, that of low frequency magnetic fields generated by the electromagnets used on the CRT (Cathode Ray Tube) screen.

Computer monitors generate these low frequency magnetic fields emanating in all directions from its position. More important to us is in the relative close proximity we keep ourselves to the monitor to read the screen and use the computer. Now we have a concern.

Excessive ELF fields emitted by computer monitors is an industry wide problem, as virtually all CRT computer monitors emit excessive ELF unless specifically stated otherwise. MacWorld Magazine (7/90 issue) did ELF studies on 10 popular computer monitors. All of the monitors tested emitted excessive ELF at close range. The only recommendation that they or I can offered you at this time is to increase the distance between you and the monitor. A working distance of two feet is recommended. Below are the results I obtained when I checked the ELF output of one computer monitor I use in my home.

TALE OF THE TAPE

The ELF field propagates from all points around the monitor, not just from the front screen. This fact becomes important in offices where computer terminals are in close proximity to one another because operators can be exposed not only from their own monitor but also from a neighbor's monitor.

It's important to realize that the ELF field given off will vary somewhat from monitor to monitor. These are the measurements of the 60 Hz ELF field I read from my 1084 Amiga monitor. My readings are given in milligauss. Magnetic field strength is measured in gauss. This unit of measure is too large for our purposes. We use 1 milligauss which is 1/1000 of a gauss.

DISTANCE	FRONT	L-SIDE	R-SIDE	BACK	TOP	BOTTOM
0"	78	97	90	125	270	N/C
4"	24	14	16	37	65	N/C
12"	5	1.5	1.5	8	9	N/C
24"	< 1	< 1	< 1	3	1.5	N/C

As you can see, the ELF field strength drops off dramatically with distance from the monitor. I could not check the ELF radiating from the bottom of the monitor because of the way it is situated in my work space.

SHIELDING

It would be nice if we could purchase a shield for our monitors, similar to the anti-glare shield on the market. Unfortunately none exist. Be very careful, as there are some anti-glare screens on the market that make a claim of blocking the electric and magnetic fields given off from the monitor. First, electric fields, as far as I know, have not been reported to have any negative impact upon health. Second, the magnetic field these screens claim to block are in fact the high frequency fields generated by the CRT. These high frequency magnetic fields have not been shown to have a negative impact on health. These screens have no impact on the low frequency (60 Hz) magnetic fields that I am talking about.

There is no easy way to shield the monitor to reduce the propagating ELF field. I have tried a number of different methods, and none had any appreciable impact on the ELF. The best recommendation is to keep the monitor 18-24 inches away from yourself.

Another possibility is to use an alternative type computer monitor. LCD (liquid crystal display) and plasma display screens do not emit ELF fields. Their drawbacks are higher cost and lower resolution.

PRECAUTIONS AROUND THE HOME

There are numerous other sources of ELF around the typical home. Before I discuss these other areas I would first like to explain dose-rate. An appliance in the home may generate a very strong ELF field, but if the appliance is only used a short time its risk factor is probably low. Note the word "probably" in the last sentence. Currently, exact data on short-term high strength fields hasn't been gathered. Electric razors fall into this category. Line operated (plugged into a wall socket rather than battery powered) razors do produce extremely strong ELF fields, and are held in very close proximity to the body, but because they are only used a short time, the total exposure or dose is small and they are probably safe.

In contrast to the electric razor is the electric blanket. Here we have a much lower ELF field strength but a much longer exposure.

Dr. Nancy Wertheimer, who first published the epidemiological study showing a correlation between 60 Hz power lines and increased incident of childhood cancer in this country, has also performed similar research on users of electric blankets. She has found that there is a higher incident of miscarriage among pregnant women who use electric blankets as compared to pregnant women who do not.

For users of electric blankets the following recommendations can be made. One, switch to ordinary blankets. If you like an electric blanket, use it to heat your bed before going to sleep, but unplug the blanket before you actually get into bed. It is not sufficient to just turn off the blanket because many blankets still produce the ELF field as long as they're plugged into the socket.

It's impossible for me to state what is a safe long-term dose-rate because it hasn't been established. Effects have been reported at dose-rates as low as 1.2 to 3 milligauss. So I would venture to say to try to limit long-term exposure of ELF to 1 milligauss or less.

TELEVISION

Television sets fall into the same category as our computer monitors. And like our monitors they produce a field that propagates around the entire set. The ELF field will propagate through standard building material such as wood and plaster. If a TV set is placed against a wall, the ELF will propagate through into the adjoining room, it becomes important not to place a bed against such an adjoining wall opposite a TV set.

VDT

VDT stands for video display terminal. This is simply a computer's monitor. Although we have discussed prominent health hazards stemming from ELF fields already, I wish to make an additional report. There have been numerous reports from female computer operators of cluster miscarriages. The word cluster refers to a greater-than-average incident of miscarriages among a group of women. The latest study on cluster miscarriages was performed in 1988 by doctors Marilyn Goldhaber, Micheal Polen and Robert Hiat of the Kaiser Permanente Health Group in Oakland, California. The study involved 1,583 pregnant women. The results of the study showed that female workers who used computers more than 20 hours a week had double the miscarriage rate as compared to female workers who did similar work without computers.

What this study didn't take into account, but which I'm sure will be studied in the near future, is the incident of malformations and cancers in the children born to the women who used the computers as compared to the children of the women who didn't. If we extrapolate the information from David Savitz' 1990 study (under Evidence paragraph) we may see another side to the problem.

FLUORESCENT LIGHTS

Fluorescent lights are much more efficient (more light per electrical watt) than ordinary incandescent bulbs. Because of this, fluorescents have become the standard lighting system used for most commercial office and industrial lighting. However, fluorescent lights require a ballast transformer that generates an ELF field. If you're using a small fluorescent lamp as a desk light you may want to consider switching to an incandescent lamp which generates virtually no ELF.

ELECTRIC CLOCKS

Small electric clocks that are plugged into a wall socket also produce an ELF field from the small internal electric motor. If this clock is an alarm clock that lies close to the sleeper's head it could be giving a significant ELF dose during your nighttime sleep. The recommendation would be to move the clock a significant distance away, or purchase either a battery-powered clock or a digital clock that produces a negligible field.

HAIR DRYERS

Hair dryers fall in the same category as electric shavers; short-term, high field strength exposure. These are probably safe for most people. Notable exceptions are for people who use these in their occupation, such as hair stylists and hairdressers.

ELECTRIC HEATERS

Electric baseboard heaters are another potential problem appliance. Recommendation is to stay a minimum of 4 feet distance from heater.

THE BOTTOM LINE ON ELF

The controversy still rages as to the impact and extent of ELF fields on human health. There is sufficient evidence to take a conservative view on the amount of ELF we allow ourselves to be exposed to. I would try to limit long term exposure to one milligauss or less. Of course it's difficult to know what your ELF exposure level is without a milligauss meter (device used to measure ELF). The ELF monitor described in this article senses the 60 Hz electromagnetic field from any appliance and indicates whether the radiation level is safe or not. The warning trip point of the circuit is about 1.5 to 2.5 milligauss. Using the meter around your home, apartment or work space will enable you to identify potentially hazardous ELF fields. Once identified, you can implement corrective action. If you decide not to build the meter, the advice is to follow the precautions outlined above. These will help to reduce your ELF exposure.

THE ELF MONITOR

The key to the ELF monitor is the sensor; a telephone pickup coil normally used to record phone conversations. The pickup coil detects 60 Hz electromagnetic fields and produces a voltage in proportion to their strength. This simple sensor doesn't have the sensitivity of more expensive ones, but it is sufficient to build a simple low cost go/no-go ELF monitor.

As shown in the schematic diagram of **Figure 5-1**, the circuit uses a dual biFET op-amp. A germanium diode, D1, in the feedback loop of the first stage provides nonlinear feedback. The diode allows the op-amp to amplify and rectify small signals from the sensor. When there is insufficient output voltage from the op-amp to drive the diode into conduction, the feedback loop is open and the op-amp operates at its full voltage gain. At this point, only a small voltage from the ELF sensor is required to produce a large output. The large output drives the diode into conduction, at which point resistor R1 determines the amplifier's gain. In practice, the millivolt-level AC signal from the sensor is put through a half-wave rectifying op-amp, where D1 compensates for the voltage drop across diode D2. The second half of the op-amp then provides additional amplification of the signal so that it is sufficient to drive the two LM339 comparators, which are used as display drivers.

We chose to use germanium diodes for D1, D2 and D3 since these have a lower voltage drop (about 0.3V) than silicon diodes (about 0.7V) and extract superior performance from the op-amps. The PC mounted potentiometer (R5) allows the unit to be calibrated (as described later). Capacitor C1 blocks any DC that may flow from resistor R2 through the coil.

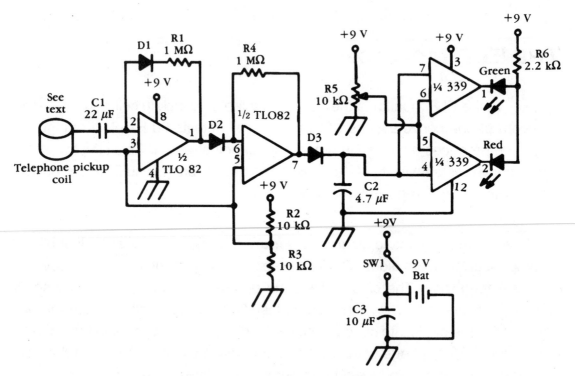

Figure 5-1. The schematic of the ELF monitor.

CONSTRUCTION

You must use a nonmetallic, nonconducting case for the project; a metal case will impede the ELF measurements. A plastic project case works well(**Figure 5-2**). The project is simple enough to use point-to-point wiring if you don't want to make the PC board. If you decide to use point-to-point wiring, keep all lead lengths as short as possible to minimize pickup in the wiring.

Once the main circuit is completed, you must connect the sensor pickup coil. The coil has shielded wire protruding from its side. Cut this cable, leaving about 3 inches attached to the coil. Split and strip the wires in the cable. Attach the pickup coil to the front of the case using epoxy or hot glue. Drill holes in the top of the case to hold the two LEDs and the power switch. The project case we used in the prototype has a battery compartment to house the 9V battery.

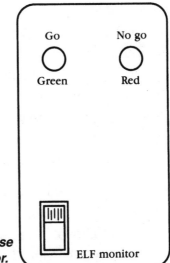

Figure 5-2. The case of the ELF monitor.

CALIBRATION

Calibration must take place in an area that is relatively free of 60 Hz radiation. Turn on the monitor and adjust R5 so that the green LED is just about to turn off and the red LED is just about to turn on. This is the proper operating point. If the environment you're working in has a significant ELF background you may have to take the unit to an isolated area to calibrate it. You can check your environment by calibrating the unit in a metal enclosure. If you own a metal desk, try placing the unit in a drawer and adjusting R5. If the red LED turns on when it is removed from the metal enclosure and the green LED turns on when you place the unit back inside, you have a significant ELF background. The monitor quickly detects an ELF field; however, it is a little sluggish (typically, there is a delay of around 1 second) in responding once the field is removed.

USING THE ELF MONITOR

The ELF monitor will accurately measure the 60 Hz magnetic field from any appliance. To test the unit, turn on a television, starting form approximately 2 feet away, and slowly bring the unit closer to the set. Once you are close enough, the green LED will turn off and the red LED will turn on. As you walk around checking various appliances, you'll probably find you can lower your ELF exposure just by rearrangement. For instance, one of my computers uses an external power supply that emits strong ELF radiation. I simply moved the power supply away from my work space. The ELF monitor can also detect a static magnetic field, when it is moved into or out of the field. If the monitor remains fixed in the magnetic field for a short time the green LED will come back on.

When measuring the ELF from a television or computer monitor you will notice the reading varies by a few milligauss, changing from low to high to low on a slow moving sine curve. This, I believe, is caused by the field phase interaction of the electromagnetic coils on the CRT tube. This effect isn't seen on transformer coils, electric motors or anything else; I have checked. This also isn't seen when using the higher power setting on the ELF monitor.

When measuring an unknown quantity you should start with the higher ranges and work down.

TROUBLESHOOTING

If you notice when switching to a higher scale that the meter reading is too high you probably have stray pickup in the circuit leads. Check the leads from rotary switch S2 to circuit; these wires are sensitive to pickup. Try making the leads as short as possible or use shielded wires.

Just in case you ask! I checked to see if the meter movement of the circuit would cause any feedback into the ELF sensor because they're in so close proximity to one another. There was none.

PARTS LIST

---◆---

ITEM	
B1	9V Battery
C1	.22 µF Mylar Capacitor
C2	4.7 µF, 16WVDC Electrolytic Capacitor
C3	10 µF, 16WVDC Electrolytic Capacitor
D1-D3	1N34A Germanium Diodes
L1	Telephone Pickup Coil
LED1	Green Subminiature LED
LED2	Red Subminiature LED
R1, R4	1Megohm Resistors
R2, R3	10 Kohm Resistors
R5	10 Kohm PC Mounted Micropotentiometer
R6	2.2 Kohm

All fixed resistors are 1/4 W, 5% tolerance

---◆---

S1	SPST Toggle Switch
U1	TLO82 Dual BiFET Op-Amp
U2	LM339 Quad Comparator
Misc	PC Board, Plastic Case, 9V Battery Clip, LED Holders, Solder, etc.

CHAPTER

◆

HYDROPHONE

Aquatic sea life produce sounds as diverse and intriguing as those of land animals. Although there are a few audio cassettes available of recorded whale songs and dolphin sounds, for the most part these underwater sounds are left unheard.

Scuba divers I am told hear a few muffled underwater sounds, but I personally never have heard any aquatic life on any of my dives. To hear underwater sounds accurately we need a device called a hydrophone. A hydrophone is designed to receive underwater vibrations and convert them into weak electrical signals. Essentially a hydrophone is an underwater microphone. The weak electrical signals can be amplified, as with a standard microphone, and listened to from a loudspeaker.

The most obvious use for the hydrophone is to listen to fish and other aquatic life like whales and dolphins. Other applications range from monitoring the turbulence created by boat and ship propellers to a splash monitor for swimming pools.

Fortunately for our project we do not have to construct the hydrophone itself; they are available from Edmund Scientific, see suppliers index in Appendix A. We are constructing a suitable audio amplifier. By connecting the hydrophone to a suitable amplifier we can listen in on underwater sounds.

HYDROPHONE SPECIFICATIONS

The hydrophone (**Figure 6-1**) can be used in either fresh or salt water. It has a operating depth of 300 feet. The frequency response of the hydrophone is 10 to 6000 Hz. Rolloff is greater than 12 dB per octave above 7000 Hz. This frequency response is fine, as most aquatic life produces sound in the low frequencies.

Figure 6-1. The hydrophone.

AUDIO AMPLIFIER

The amplifier for the hydrophone is made up of four IC modules (see **Figure 6-2**). The signal from the hydrophone is pre-amplified by IC1, configured as a 15X amplifier. The amplified signal passes through IC2, a 60 Hz notch filter and IC3, a low-pass filter with a 7000 Hz frequency cutoff. The filtered signal is amplified by IC4, an LM386 audio amplifier to the speaker SPK1. Potentiometer R12 controls the gain of the LM386 amplifier.

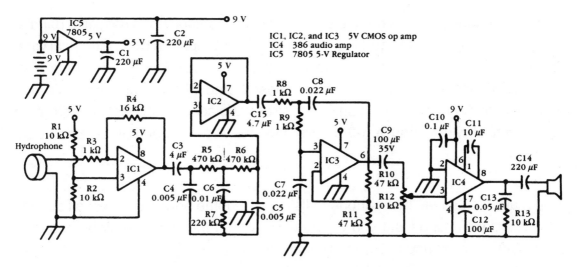

Figure 6-2. The schematic of the hydrophone.

There is nothing critical on the construction. The most difficult part is securing the speaker to the housing. I placed speaker cloth and the speaker on the opening in the housing and used hot glue to secure everything. If you don't have a hot glue gun, I'm sure epoxy will work just as well.

The hydrophone comes with attached leads that are connected with shielded cable to a 1/8 inch phono plug. The wire coming out of the hydrophone is thin. Carefully strip away the insulation and separate the wires. Solder one wire from the hydrophone to the shield of the cable; this should be the ground on the phono plug. When the wires are soldered, secure the connections with electrical tape. Then waterproof the connection with silicone rubber sealant, available from tropical fish stores to fix leaks in aquariums.

USE

To test the unit, turn R12 volume control for maximum gain. Pass your finger across the surface of the hydrophone; you should hear a loud scraping sound from the speaker.

The hydrophone can be mounted various ways depending upon use; on a handle, pole or secondary cable line. Do not support the hydrophone by its electrical wire. Instead use a secondary cable, and wrap the cable around the hydrophone. Secure the cable to the hydrophone using a hose clamp. Wrap or tape the audio wire to the secondary cable line. Insure that stress is placed on the cable and not on the electrical wires. **Figure 6-3** shows how this above setup can be used off a boat.

You may also use the hydrophone to listen in to home aquariums. Not all aquatic animals make sounds, but you will need to turn off the filter and air pumps from drowning out all sounds in order to hear anything.

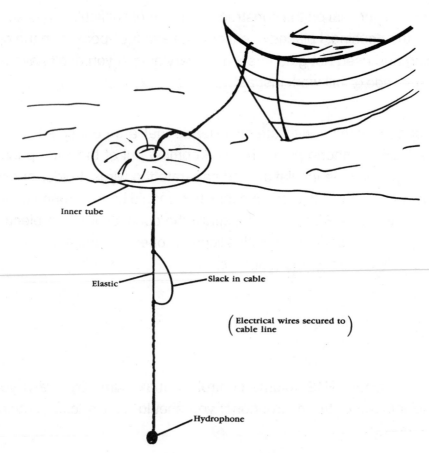

Inner tube

Elastic   Slack in cable

(Electrical wires secured to cable line)

Hydrophone

Figure 6-3. The hydrophone being used off of a boat.

PARTS LIST

◆

ITEM			SOURCE
Hydrophone	# C41,759		Edmund Scientific
IC1, IC2, IC3	5V CMOS Op-Amp		Images Company
IC5	7805 Voltage Reg.	#276-1770	Radio Shack
IC4	LM386 Audio-Amp	#276-1731	Radio Shack
R1, R2	10K	#271-034	Radio Shack
R3, R8, R9	1K	#271-023	Radio Shack
R4	15K	#271-036	Radio Shack
R5, R6	470K	#271-053	Radio Shack
R7	220K	#271-049	Radio Shack
R10, R11	47K	#271-042	Radio Shack
R13	10 ohm	#271-001	Radio Shack
R12	10K Potentiometer	#271-1715	Radio Shack
C1, C2, C14	220 μF	#271-1029	Radio Shack
C3	47 μF	#272-1027	Radio Shack
C4, C5	.0047	#272-130	Radio Shack
C6	.01 μF	#272-1065	Radio Shack
C7, C8	.022 μF	#272-1066	Radio Shack
C9, C12	100 μF	#272-1028	Radio Shack
C10	.1 μF	#272-1432	Radio Shack
C11	10 μF	#272-1025	Radio Shack
C13	.047 μF	#272-1068	Radio Shack
C15	4.7 μF	#272-1024	Radio Shack
SPK1	2 1/4" Speaker	#40-246	Radio Shack
Misc	Housing, 9V Battery, Cap & Holder, Knob, PC-Board, Shielded Cable, 1/8 inch Phono Plug and Jacket		

◆

CHAPTER

◆

GEIGER
COUNTER

We are surrounded by energy that we can not observe with our senses. For instance, turn on a radio and you can listen to broadcasted electromagnetic signals that have been passing through you and your home completely unnoticed. You never sense this electromagnetic radiation directly because you can't see, hear, feel or taste it with any of your senses. The same is true of radioactivity. To detect radioactivity we will build an instrument called a Geiger counter that will give a visual and audible indication of local radioactivity.

RADIOACTIVITY

Simply put, radioactivity is the spontaneous emission of energy from the atomic nucleus of certain elements. The energy emitted can take the form of particles as in alpha and beta particles or electromagnetic energy as in gamma rays.

In general, people are concerned with radioactivity and ionizing radiation. This is due to its relationship to atomic weapons and nuclear power plants. There is also a legitimate concern over long-lived waste products from nuclear power plants and their potential for ecological damage. However legitimate this concern, it is usually overstated due to bad press and grade B science fiction movies.

When a nuclear power plant has an accident, such as at Three Mile Island, the media and a few anti-nuclear groups go into a feeding frenzy. This fear mongering pushes public fears to the point where it becomes so overwhelming that nuclear power plants are forced to shut down before they have started producing power. A case in point is the Lilco nuclear power station in Long Island. New York, shut down by Gov. Mario Cuomo. This attitude overlooks the medical, scientific and economic benefits this technology has offered humanity.

Please Note: I am not saying that anyone should treat atomic radiation or nuclear materials lightly. You should not; but it is not a technological monster either.

BACKGROUND RADIATION

Perhaps we can put radioactivity into better perspective by looking at some natural sources. To begin with, life on this planet has always been exposed to an environment in which cosmic rays, along with naturally occurring radioactive materials in the soil (uranium-238 and thorium-232) and food (potassium-40 and carbon-14), maintain a continuous background level of radiation. In essence we have evolved in this radioactive environment where millions of gamma rays pass through each individual on earth every hour.

The background radiation will cause the Geiger counter to click about 12-14 times a minute. This is normal. As stated there are millions of gamma rays penetrating every one of us every hour, so naturally not all the radiation passing through the GM tube is detected.

The Geiger counter we will construct can be used to detect nuclear radiation or contamination in your home or local area, prospect for uranium, measure the background radiation, and sometimes even detect solar flares.

HISTORY

Radioactivity was discovered in 1896 by a French scientist Henri Becquerel. He found that pitchblende ore (which contains uranium-238) fogged photographic plates even though the plates were kept completely covered and unexposed to light. Today we know that this radiation consisted of alpha *(a)* particles. There are two other types of radiation emitted from radioactive materials: beta *(B)* and gamma *(Y)*. The term radioactivity was coined by Pierre and Marie Curie. They also proved that radioactivity is an atomic property and not a chemical one.

Alpha particles are helium-4 nucleus consisting of two protons and two neutrons. When an atomic nucleus emits an alpha particle it changes into another nuclide with an atomic number of two units less and a mass number of four units less. For instance when uranium-238 emits an alpha particle it transmutes to thorium-234. Alpha particles do not penetrate very far. They can be stopped by a few sheets of paper or a few inches of air.

Beta particles are either electrons or positrons (positive electrons). Beta particles are more penetrating than alpha particles, but they can be stopped by thin metal sheets such as aluminum or a few feet of air.

Gamma rays have deep penetration, they can go through several inches of metal. Gamma rays are photons of energy (quanta) emitted from excited atoms. They are not associated with any particle. When an atom such as uranium-238 emits an alpha particle it becomes thorium-234. The thorium atom at this point has excess energy. It is said to be in an excited state. By emitting a gamma ray it drops to its ground state (unexcited).

MEASUREMENT OF RADIOACTIVITY

There are a number of ways to measure radioactivity such as scintillation, gas ionization, pn junctions and autoradiography.

The Geiger-Mueller tube works on the gas ionization principal. Surprisingly, the Geiger-Mueller tube of today looks pretty much the same as the original tube, except for the fact that it is miniaturized. The GM tube we are using in this project detects alpha, beta and gamma rays.

In 1920 it was decided that the amount of radiation given off by one gram of radium be named the Curie, in honor of the discoverers of radium. This unit of measure is equal to 37,000,000,000 (3.7×10^{10}) atomic breakdowns per second. The Curie is a rather large number and many radioactive sources are measured in millicuries (1/1000 of a Curie) or microcuries (1/1,000,000 of a Curie). A micro-curie (uci) is equal to 3.7×10^4 or 37,000 breakdowns a second.

GEIGER TUBE

Figure 7-1 shows the basic operating principal of the Geiger-Mueller tube. The tube is constructed with a cylindrical electrode (cathode) surrounding a center electrode (anode). The tube is evacuated and filled with a Neon and Halogen gas mixture. A voltage potential of 500 volts is applied across the tube, through a 10 megohm current-limiting resistor R1. The detection of radiation relies upon its ability to ionize the gas in the GM tube. The tube has an extremely high resistance when it is not in the process of detecting radioactivity. When an atom of the gas is ionized by the passage of radiation, the free electron and positive ionized atom created move rapidly towards the two electrodes in the GM tube. In doing so they collide with and ionize other gas atoms which creates an small avalanche effect. This ionization drops the resistance of the tube, allowing a sudden surge of elec-

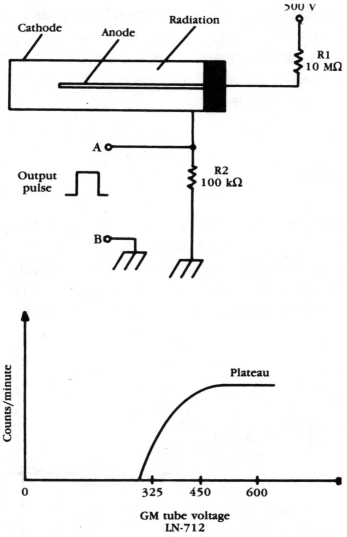

Figure 7-1. The GM tube and operation chart.

tric current that creates a voltage across the resistor R2 that we see as a pulse. The Halogen gas quickly quenches the ionization, thus returning the GM tube to its high initial resistance and its ability to detect another particle.

The number of pulses per minute the GM tube responds to rises with the voltage potential across its electrodes. By increasing the voltage we reached a plateau where the count rate stays pretty constant. The plateau range for our GM tube lies between 400 and 600 volts with a recommended operating voltage of 500V. If one applies too much excess voltage to the tube it may damage the tube. In this case, when the tube detects a radioactive particle with too high a voltage, the avalanche created will not be quenched, putting the tube in a state of continuous discharge that can damage the tube.

THE CIRCUIT

The circuit is shown in **Figure 7-2**. IC2 is a 555 timer set in astable mode. The signal from IC2 is presented to three gates on the 4049 IC1. The 4049 inverts the signal to give an optimum pulse width that switches Q1 on and off. The MOSFET (Q1) in turn switches the current to a step-up transformer TR1. The stepped up voltage from TR1 first passes through a voltage doubler, the output voltage from

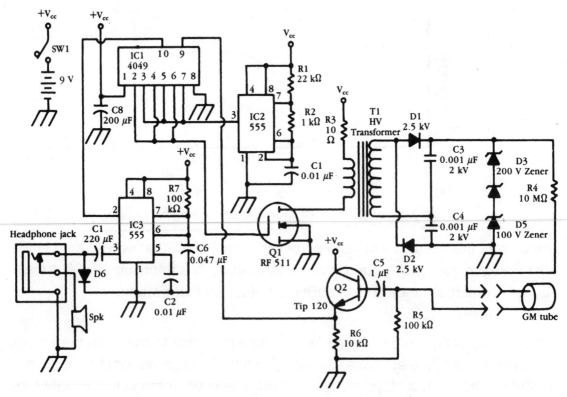

Figure 7-2. The schematic of the Geiger counter.

this section being approximately 600-700 volts. Three zener diodes (D3, D4 and D5) are place across the output of the voltage doubler to clamp and regulate the voltage to 500V. The 500V is connected to the anode on the GM tube through a 10 megohm resistor. The resistor limits the current through the GM tube and allows the detection ionization to be quenched. On the cathode side of the tube is a 100K connected to ground. When the GM tube detects a particle, a voltage pulse from the 100K resistor is amplified and clamped to Vcc via Q2 an NPN Darlington transistor. The signal from Q2 is inverted by a gate on IC1, where it acts as a trigger signal to IC3. IC3 is another 555 timer configured to monostable mode, that stretches each pulse received. The output of IC3 via pin 3 flashes the LED and provides a click into either the speaker or headphones.

The circuit is powered by a 9V alkaline battery and draws about 28 mA when not detecting.

CONSTRUCTION

There is nothing critical about the circuit; it can be hand wired (see **Figure 7-3**). Of course using a PC board for the project will make construction that much easier. Any plastic enclosure large enough to hold the circuit board and other components is fine.

The GM tube is delicate and should be handled carefully. On the front end of the tube is a thin mica window that allows alpha and beta rays to penetrate and be detected. Be careful, because this window is easy to break and would render the tube useless. It's a good idea to enclose the GM tube in its own housing. A microphone case

Figure 7-3. Inside view of the Geiger counter.

is an excellent housing. **Figure 7-4** shows the Geiger counter. Images Company sells a microphone for $4.95 that looks as though it will work. I used a plastic coin tube for my GM tube housing. I cushioned the tube using soft foam rubber around the diameter (see **Figure 7-5**). I cut some foam the diameter of the tube and stuffed it into the tube to take up the dead space at the bottom. (NOTE: if you use conductive foam do not allow the foam to touch both electrodes on the GM tube.

Figure 7-4. The Geiger counter.

Figure 7-5. The GM tube probe.

This will put a short across the GM tube rendering the unit inoperative.) On the top cover of the tube I drilled a number of small holes to allow alpha and beta particle through to the mica window unimpeded. I soldered about three feet of two-conductor shielded wire to the bottom leads on the tube, threaded through a drilled hole in the bottom of the coin tube. At the opposite end of the shielded wire I soldered a 1/8 inch plug that connects the tube to the main circuit. This allows you to remove the GM tube from the case for storage. I also secured a small amount of velcro to the side of the case and coin tube to secure the GM housing to the case.

RADIOACTIVE SOURCES

An easy source of a radioactive material is an ionization-type smoke detector. These smoke detectors contain a small amount (approximately one microcurie) of americium-241 inside. Americium is a strong alpha particle source. To use the americium as a source you must remove it from the smoke detector. The alpha particles only travel an inch or so through the air so you have to get pretty close to it with the GM tube to detect anything.

To get to this radioactive source, first remove the plastic top of the smoke detector. The americium is inside a small ventilated metal can. Remove the can with pliers. The americium is embedded into a metal plate underneath the can. Remove the entire plate with the americium from the detector. Leave the radioactive material attached to the plate, and use the material on the plate as a source.

By bringing the GM tube 1/2 inch away from the source the clicking will become furious.

A more reliable source of nuclear materials is the Nucleous Company, see parts list. This company sells calibrated and uncalibrated radioactive sources for students, schools and industry. I purchased a cesium-137 gamma ray source for $25.00. This material is rated at 5 microcuries with a 30-year half life. It was pretty enlightening when I placed a 3/4 inch solid block of aluminum in front of the cesium-137, and found there was no noticeable decrease in radiation.

CHECK-OUT & TROUBLESHOOTING

Before plugging in the GM tube it's a good idea to see if the circuit is functioning properly. Turn on the circuit; you should measure approximately 500V from the junction of D3 and C3 with respect to ground. If you're not getting a proper reading, first check the zener diodes to make sure you have them facing the right direction. Next, check pin 3 of IC2; you should read a 5000 Hz square wave pulse. Trace this pulse through IC1 which inverts this signal going to Q1. Q1 powers the step-up transformer TR1. The output of TR1 is connected to a voltage doubler regulated by the zener diodes. This is the main power supply for the GM tube.

When the circuit checks out, turn it off and plug in the GM tube, then turn it on again. In my area I read approximately eleven pulses per minute from background radiation. As each particle is detected the speaker will click and the LED will flash.

If you have acquired some radioactive material, bring the GM tube close to it to test for activity.

OTHER GEIGER TUBES

The circuit is designed around the LN-712 Geiger tube. It is possible to use other Geiger tubes. The power supply can power any tube up to about 700 volts. You can change the voltage output of the circuit by changing or removing the zener diodes across the output. By doing so you can vary the voltage output up or down.

If for instance you have a tube that requires 350 volts, simply remove one of the 200 volt zener diodes and use a jumper wire between the two open terminals and you have a regulated 350 volt supply. If you need 600 volts switch out the 100 volt zener diode with a 200 volt zener and you have a regulated 600 volt supply.

Other tubes may require a different current limiting resistor. This becomes a simple matter of removing R4 and replacing it with the proper current limiting resistor for that particular tube.

INVERSE SQUARE LAW

All radiation follows the inverse square law, which states that the intensity of the radiation is proportional to the square of the distance from its source. Put another way; if the distance from the source is 2 then the intensity is 1/4 (if the distance is 3 then the intensity is 1/9) of the intensity of the radiation at a distance of 1. What this means for our small radiation sources is that the intensity falls off pretty dramatically with distance. Radiation will deviate (greater losses) from the inverse square law due to air absorption and scattering (alpha and beta). Another factor that weighs heavily is the area of the detector exposed to the radiation (gamma).

DETECTING SOLAR FLARES

As stated previously part of the natural background radiation on earth is due to cosmic rays. The origin of cosmic rays can be galactic, entering our solar system from all directions. Or they may be solar cosmic rays emitted by the sun.

The energy emitted from the sun is more or less continuous. However, when a solar flare occurs there is a rapid eruption of x-ray and UV energy from 10 to 100 times the normal level. This burst of energy may be detected on earth using a simple GM counter. To do this experiment one must make prolonged recordings and subtract the background radiation from the readings. This will identify bursts of radioactivity that can be correlated to current solar activity. (see NOAA)

PARTS LIST

	ITEM		SOURCE
IC1	4049 Hex Buffer	#276-2449	Radio SHack
IC2 & IC3	555 Timer	#276-1723	Radio Shack
Q1	IRF511 MOSFET	#276-2072	Radio Shack
Q2	TIP-120 NPN	#276-2068	Radio Shack
R1	22K	#271-1339	Radio Shack
R2	1K	#271-1321	Radio Shack
R3	10 ohm	#271-1301	Radio Shack
R4	10 Meg	#271-1365	Radio Shack
R5 & R7	100K	#271-1347	Radio Shack
R6	10K	#271-1335	Radio Shack
C1 & C2	.01 µF	#272-1065	Radio Shack
C3 & C4	.01 µF 2 KV Cap	#272-160	Radio Shack
C5	1 µF	#272-996	Radio Shack
C6	.047 µF	#272-1068	Radio Shack
C7 & C8	220 µF	#276-956	Radio Shack
D1 & D2	1 KV Diode	#276-1114	Radio Shack
D6	Red LED	#276-044	Radio Shack
Spk	8 ohm 2.25" Speaker	#40-246	Radio Shack
D3 & D4	1N5388B 200V Zener		Newark Electronics
D5	1N5271B 100V Zener		Newark Electronics
TR1	C-2B Transformer		Allegro Electronics
GM Tube	LN-712 GM Tube		Images Company
PC Board	PCB-GM1		Images Company

CHAPTER

---◆---

MHD
GENERATOR

A Magnetohydrodynamic (MHD) generator produces electrical power. As in a conventional generator, it produces power by moving a conductor through a magnetic field. The moving conductor in a standard generator is a coil of copper wire. Unlike a standard electrical generator, the MHD contains no moving parts. In the MHD, the conductor is a fast moving hot plasma gas.

MHD BASICS

Figure 8-1 illustrates the basic operation of the MHD generator. The high temperature electrically conductive gas flows past a transverse magnetic field. An electric field is generated perpendicular to the direction of gas flow and the magnetic field. The electric field generated is directly proportional to the speed of the gas, its electrical conductivity and the magnetic flux density. Electricity can be siphoned off with electrodes placed in contact with the flowing plasma gas.

The MHD generators require a strong magnetic field. In order to make MHD generators a practical energy supply, superconductive magnets must be used.

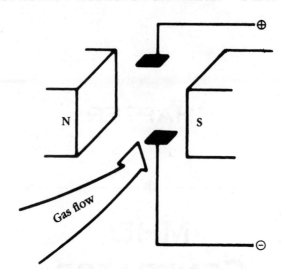

Figure 8-1. The basic MHD generator.

MAKING A PLASMA

The plasma in the MHD is created by a process called thermal ionization, where the temperature of the gas is raised to the point so that the electrons are no longer bound to the atoms of gas. These free electrons make the plasma gas electrically conductive.

To create such a plasma through thermal ionization alone requires extraordinary high temperatures. The gas temperature can be lowered significantly by seeding the gas with an alkali metal, such as potassium nitrate. The alkali metal ionizes easily at lower temperatures.

In our model, the gas is continuously seeded with potassium nitrate, making the gas electrically conductive at lower temperatures.

ADVANTAGES OF MHD GENERATORS

Conventional coal-fired generators achieve an maximum efficiency of about 35%. MHD generators have the potential to reach 50% or 60% efficiency. The higher efficiency is due to recycling the energy from the hot plasma gas to standard steam turbines. After the plasma gas passes through the MHD generator, it is still hot enough to boil water to drive steam turbines that produce additional power.

MHD generators are also ecologically sound. Coal with a high sulfur content can be used in the MHD without polluting the atmosphere.

MHD MODEL

The table-top model MHD generator is pretty easy to build. There are numerous heat and conductivity losses due to its simple construction. You should look at this as an opportunity to vastly improve upon its design and consequently its power output.

Figure 8-2 illustrates the construction. Two large ceramic magnets are self-supported using three steel plates, see parts list. The self-supporting magnets form a chamber where the electrodes are positioned. A small base-board supports six wire electrodes. The electrode height is matched to the burner height of the torch used. The electrodes are wired as shown in **Figure 8-3**. The use of segmented electrodes produces a greater electric output than a single large electrode.

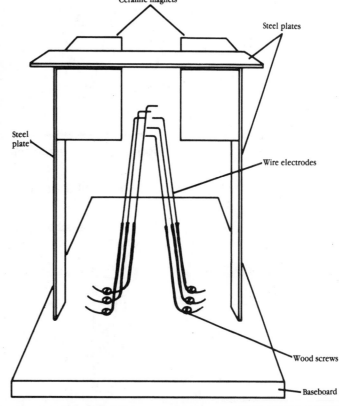

Figure 8-2a. The MHD generator.

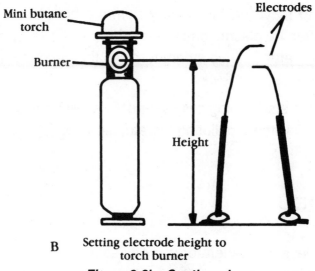

B   Setting electrode height to
torch burner

Figure 8-2b. Continued.

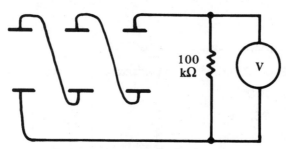

Figure 8-3. The electrode wiring.

Figure 8-4 shows a simple seeder unit that seeds the gas with potassium nitrate.

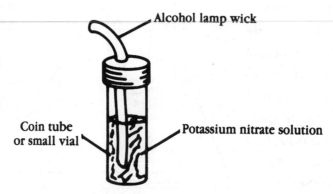

Figure 8-4. The seeder unit.

OPERATION

Operating the MHD is simple. The wick of the seed unit is placed so that it touches the flame of the miniature torch. The flame of the torch is made as large as possible and directed in between the ceramic magnets hitting the electrodes.

Set your VOM to its most sensitive scale for volts. The prototype MHD produced about .1 volts.

IMPROVING THE MHD GENERATOR

The model MHD generator is working with a mediocre temperature gas. A hotter and larger flame from an oxy-propane or any number of other torches will improve conductivity of the gas and thus performance.

The ceramic magnets produce about 4000 gauss, pretty weak by MHD standards. Replacing the ceramic magnets with stronger electromagnets will increase electrical power output.

Finally, the electrodes should be replaced with a tungsten or nichromium wire. The standard wire used in the prototype wears out pretty fast. The high temperature wires will last substantially longer. You may be able to use the tungsten wire filament from burned out incandescent lamps.

By making these changes, the MHD power produced will become more pronounced. In the prototype, thermionic (gaseous thermocouple) effects contribute more electrical power than the MHD effect.

LIQUID METAL MHD

We are using a gaseous conductor flowing past a magnetic field to induce voltage. A conductive liquid could also be used to achieve the same effect. A liquid metal such as mercury has been used in an MHD configuration to produce power.

MHD PROPULSION

Since the movie *The Hunt For Red October*, created from Tom Clancy's best selling novel, MHD propulsion systems have captured the public's attention. The MHD drive in the movie was depicted as being invisible to sonar systems.

The Navy has and is investigating MHD propulsion systems. However, they are far from being invisible at this point in time. First, the tremendous magnetic field needed to produce thrust would be relatively easy to pick up. In addition, the electric current flowing through the sea water would electrolyze it, producing a noticeable trail of gas bubbles. Although MHD propulsion has been stated to be ecologically safe, I have my doubts. The large magnetic fields and high current density flowing through the water would appear to have a negative impact on sea life. As with the MHD generators, superconductive magnets must be employed to make the drives practical.

HOW MHD PROPULSION WORKS

Figure 8-5 shows the basic MHD propulsion system. It is similar to the MHD generator. The main difference is that instead of drawing electricity via the electrodes, electricity is supplied to flow between the electrodes. The electrified seawater, being under a strong magnetic field, is propelled perpendicular to the magnetic and electric fields.

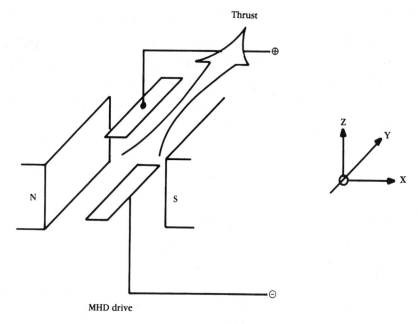

Figure 8-5. The MHD drive.

MHD PROPULSION IN SPACE

MHD propulsion systems seem to be ideally suited for interplanetary space travel. MHD rockets could not replace chemical rockets for the raw power needed to escape Earth's gravitational field because of the MHD's low specific thrust. However, once lifted in space the MHD propulsion can easily surpass and speed past chemical rockets in traveling to other planets.

The reasons are that despite their low thrust, the MHDs use very little fuel, and the exhaust velocity of the propellant is very high. Combine these two properties and you have a rocket engine that can run continuously for a long period of time that will allow the vessel to slowly gather to high speed. Remember Newton's law of motion, for every action there is an opposite and equal reaction. With chemical rockets, regardless of how powerful the thrust is, the rocket can go no faster than the exhaust velocity of its propellant.

The MHD space drive looks like the illustration in **Figure 8-5**. The difference is that an ionized gas is released in the MHD chamber. The ionized gas is accelerated due to the electric and magnetic fields and produces thrust, in the same manner as described for the sea water MHD propulsion.

ITEM	SOURCE
Large Bar Magnet 1"x 1" x 2"	Images Company
Steel Plate 1" x 4.5" x 1/16 thick	Images Company
Potassium Nitrate 1/2 oz.	Images Company
Wick	Images Company
Propane or Butane Torch	Purchase locally

CHAPTER

---◆---

PLASMA
ACOUSTICS

About 20 years ago I read an interesting article on three researchers, W. Babcock, K.L. Baker and A.G. Cattaneo, who discovered an interesting phenomena of plasma acoustics while working at the United Technology Center in California.

The researchers were able to use a high temperature plasma as a speaker. They noted that the frequency response of the plasma speaker is much better at high frequencies than low. This drawback, however, may be due to the small size of the plasma, rather than an intrinsic limitation.

Rather than try to create a high temperature plasma, we can use a lower temperature flame from a propane or butane torch that is seeded with potassium nitrate.

BASIC OPERATION

Any sound source can be used, but I'd advise using music since there are more high frequency components in it than speech. The audio source (radio, tape or CD player) must supply 10 or more watts to be heard. If necessary, feed the audio signal through a power amplifier. The more power the greater the volume. From the amplifier, the audio is fed to a step-up transformer, see **Figure 9-1**. The output of the step-up transformer is fed into two electrodes positioned in the flame. The flame is seeded with a water solution of potassium nitrate to induce low temperature ionization.

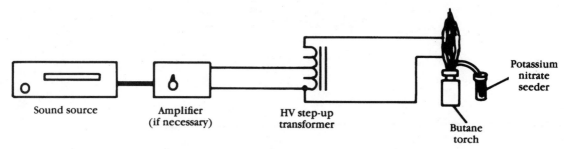

Sound source   Amplifier (if necessary)   HV step-up transformer   Butane torch   Potassium nitrate seeder

Figure 9-1. A diagram of the flame speaker components.

HOW IT WORKS

The torch flame ionizes the potassium nitrate easily, by thermal ionization. Ionization of the potassium in the flame creates free electrons and positive potassium ions. When we supply a high voltage electrical signal, the positive ions move toward the negative electrode while the free electrons move toward the positive electrode. This movement, as in the case of a audio signal, switches back and forth rapidly, causing the flame to vibrate.

Flames, like a liquid, exhibit a surface tension. A gaseous membrane is created by the difference in density and high temperature of the torch flame to the ambient air. This membrane acts as a diaphragm, compressing and rarifying the air to produce sound waves as the flame vibrates.

EXPERIMENTING WITH PLASMA ACOUSTICS

Figure 9-2 illustrates a simple set up for you to begin experimenting. The seeder I used was fashioned from an old coin tube holder, though any small container will also work. An alcohol lamp wick is used to continuously supply the potassium nitrate into the flame.

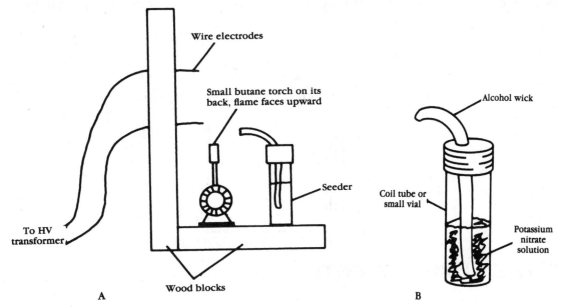

Figure 9-2. A schematic of the flame speaker.

The potassium nitrate solution strength isn't critical; dump a spoonful in a small quantity of water and it will work fine. The step-up transformer is a high voltage autotransformer from Images Co.

When running the plasma speaker, if the electrodes start arcing either move the electrodes further apart or turn down the volume on the amplifier.

Turn off all the equipment before adjusting or touching the electrodes. They are connected to the high voltage transformer and are quite capable of giving you a shock.

Iᴍᴘʀᴏᴠɪɴɢ Tʜᴇ Dᴇsɪɢɴ

The experiment can be improved by using a larger and hotter flame. Anything that would do so will increase the quality of the sound. If possible, use nichrome or tungsten wire for the electrodes. These wires will last longer in the high temperature flame. Finally, supplying a 500 VDC bias to the circuit will improve the quality of sound, see **Figure 9-3**. You may be able to use the HV section of the Geiger-Mueller Counter project to this end.

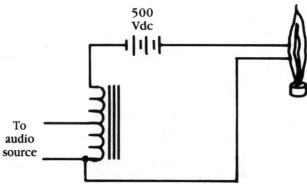

Figure 9-3. A schematic to add dc bias.

Oᴛʜᴇʀ Eхᴘᴇʀɪᴍᴇɴᴛs

A solar cell connected to an audio amplifier will output the sound signal fed into the flame when the light from the flame is directed onto the photocell by a lens, see **Figure 9-4**.

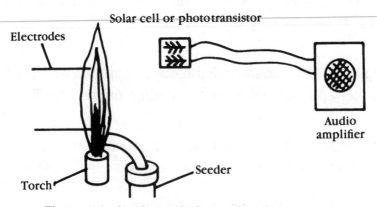

Figure 9-4. A schematic for testing light output.

Another possibility is to direct a laser beam between the electrodes in the flame. This may be a low cost way to modulate a laser with information, see **Figure 9-5**.

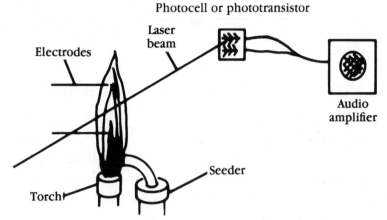

Figure 9-5. A schematic for laser modulation.

In the 1920s Lee DeForest replaced a crystal detector in a radio using a flame, see **Figure 9-6**.

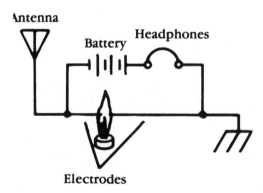

Figure 9-6. A schematic from Lee Deforest in 1920.

PARTS LIST

ITEM	SOURCE
IMT1 HV Transformer	Images Company
Potassium Nitrate 1/2 oz	Images Company
Wick	Images Company
Butane or Propane Torch	Purchase Locally

CHAPTER

♦

EXPANSION
CLOUD CHAMBER

Cloud chambers are simple devices that allow you to see trails made from atomic particles. The particular chamber we are building is a derivative of a unit described in an April 1956 *Scientific American*. The unit is updated to take advantage of newer materials. I like this cloud chamber for its simplicity and usefulness.

CONSTRUCTION

Figure 10-1 shows the basic construction. We are using two medium sized clear plastic jars. A large hole is cut into each screw-on lid. The plastic lids are fastened together using epoxy. Use a good amount of epoxy to hold these lids together or the unit will leak.

A hole is made in the side near the bottom of jars J1 & J2. The hole in J2 is fitted with a plastic elbow, secured with a water resistant cement or epoxy. On the inner end of the tube, a small rubber balloon is attached.

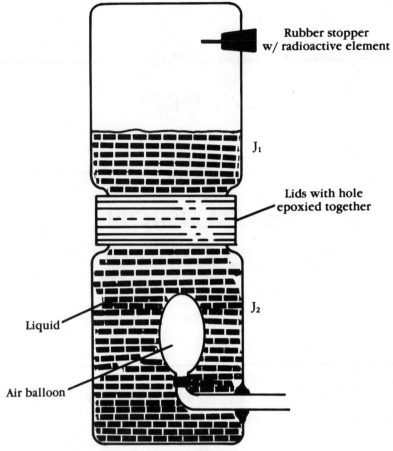

**Rubber stopper
w/ radioactive element**

J₁

**Lids with hole
epoxied together**

J₂

Liquid

Air balloon

Figure 10-1. A schematic of the cloud chamber.

Coat the threads on the jar lids with a liberal amount of an aquarium sealant before screwing on the jars. Allow sealant to cure.

Fill the jars using a funnel with a plastic tube through the hole on J1. The solution is a 1:1 mixture of alcohol and water. Add to this solution some salt and enough black India ink to turn the solution a dark black.

Finish off the chamber by gluing or securing a radioactive source to a rubber stopper. Insert the stopper into the hole in J1.

The stopper should form an airtight fit with the jar. If not, the air will escape when used rather than compress, rendering the cloud chamber useless. **Figure 10-2** shows the completed cloud chamber.

USE

Using the cloud chamber is straightforward. You inflate the balloon to raise the liquid an inch or so, using a small aquarium type air pump, or your lungs if you're healthy enough. Keep the balloon inflated a minute for the compressed air in J1 to reach room temperature. The air is then released, and the cooling effect produces clouds in the chamber that allows the tracks of atomic particles to be seen.

You will be surprised by the force required to inflate the balloon; consequently, the force generated tries to pry the jars apart. Be sure you use a good epoxy and enough of it to form a good solid seal between the lids or they will leak.

Figure 10-2. The cloud chamber.

CHAPTER

Nitinol-Shaped Memory Alloy

Shaped Memory Alloys (SMA) have some interesting properties. One property of the material is that it contracts when heated. This is analogous to the contraction of muscle tissue. Notice that this effect is the opposite of standard metals which expand when heated and contract when cooled. Using this material we can achieve electrical movement without using motors, stepper motors or solenoids.

Another property of the material is called the Shaped Memory Effect (SME). Simply put, this material will, when heated to a critical temperature, return to a predefined shape it has been trained to remember. So an object made of this metal material could be twisted, bent and folded out of shape, then heated to get the original shaped object back. It's fascinating to watch this happen. As the material is heated it quickly unbends, unfolds and untwists itself back into its original shape, kind of like self-healing.

HISTORY

In 1932, Arne Olander a Swedish researcher, discovered the Shaped Memory Effect in a gold-cadmium (Au-Cd) alloy.

In 1951 two researchers, L.C. Chang and T.H. Read, analyzed the crystal structure and changes of the Shaped Memory Effect in the Au-Cd alloy. In 1958 these two researchers made a cyclic weightlifting device to be displayed at the Brussels World Fair.

In 1961 William Beuhler working at U.S. Naval Labs, discovered SME in an alloy of titanium-nickel. At the time the Beuhler team were looking to develop a heat and corrosive resistant alloy. In any case, this alloy was by far cheaper and safer to work with than any SME alloy to date. The team named the new alloy nitinol, pronounced night-in-all. The material's name is representative of its elemental components and place of origin. The "ni" and "ti" are the atomic symbols for nickel and titanium, the "nol" stands for the "Naval Ordnance Laboratory" where it was discovered.

In the sixties and seventies other alloys were discovered that exhibited SME.

In 1985, Dr. Dai Homma of Japan's Toki Corporation announced an improved version of nitinol. This improved version of nitinol is sold in this country under the name trade name BioMetal™. References to either Flexinol, BioMetal, Muscle Wire or nitinol are to be considered one and the same.

APPLICATIONS

Many interesting applications of this material have been put forth. NASA had proposed using nitinol to make space craft antennas that would deploy when heated by the sun or a secondary heating unit.[8] More down to earth ventures are its use in eye-glass frames, dental alignment material, pumps, blood filters, solenoids and an artificial heart.

How it Works

The properties of nitinol rely upon the crystal structure of the material. The structure is sensitive to both external stress and temperature. Before we discuss the actual mechanics involved we must first define the temperature phases of the material.

Parent Phase. Material above transition temperature. Transition temperature is dependent upon the exact composition of material. For the nitinol wire we are working with, this temperature is 100-130°C (190-260°F). This temperature is where the wire contracts or returns to a previously defined shape. The crystal structure is cubic.

Martensitic Phase. Material is below transition temperature. The crystal structure is needlelike and is collected in small domains. Within each domain the crystals are aligned. The material is cool and can be bent or formed into other shapes. This external stress transforms the crystal structure of the material. It is sometimes called stress-induced martensite.

Annealing Phase. At this temperature the material will reorient its crystal structure to remember its present shape. Annealing phase for the material we are working with is 540°C.

When a cooled wire is bent or twisted, the crystal structure is transformed. If the wire is now heated above its transition temperature (Parent Phase) the crystal structure changes from needlelike to cubic. Since the cubic crystals don't fit into the same space as the needlelike crystals, they are formed under strain. To relieve this strain they move and change their positions to relieve the strain. This "lest strain" position happens to be the original shape (or the annealed shape) of the material.

Where the wire hasn't any stress-induced transformations, the crystal structure still changes but results in no net movement.[9, 10]

PROPERTIES

Nitinol metal can generate a shape-resuming force of about 22,000 pounds per square inch. We will not be working with a square inch of material; we shall use a 6 mil wire. Even so, a 6 mil wire (.006 inch dia.) can generate a contractive force of 11 ounces. If you want more pull, simply multiply the wires till you reach the contractive force you require.

The wire can contract up to 10% of its length. For longer lifetime (greater than 1,000,000 cycles), you should restrict the contraction to only 6% of its length.

The easiest way to heat the wire is by passing an electric current through it. Care should be given not to overheat the wire or its properties will degrade. The wire has a electrical resistance of a little less than one ohm per inch. Nitinol wire is supplied with crimp terminals, see **Figure 11-1**. These terminals are used to make connection to the material because the nitinol wire should not be raised to a high temperature that would be required for soldering.

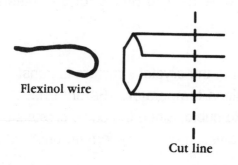

Flexinol wire

Cut line

Figure 11-1. Making crimp terminal connections.

Reaction time can be quite short, measured in milliseconds. In addition, full strength is developed at the beginning of the cycle. This is in contrast to standard solenoids which develop full strength near the end of their cycle.

The nitinol material is stronger than many steels. The 6 mil wire has a breaking strength of about 6 pounds.

WIRE DIAMETER

Sometimes nitinol wire diameter is given in micrometers. To convert a nitinol wire with a diameter of 150 micrometers to inches multiply by .00003937. Doing the multiplication we find the diameter is 6 mils English (.006 inch).

ACTIVATING NITINOL WIRE

As stated, the nitinol wire is activated by passing an electric current through the wire. The wire's resistance to the current heats the wire and causes it to contract. The volume of the wire doesn't change during contraction. So as the wire decreases in length, its diameter increases by a proportional amount keeping the volume the same. The activation temperature of the wire is 100 to 130°C (or 190-260°F).

DIRECT ELECTRIC HEATING

Nitinol wire can be activated using low voltage, such as a 9V transistor battery. A simple circuit can be constructed using a battery, switch and a small length of nitinol, see **Figure 11-2**. Care must be taken not to overheat the wire. In addition, direct electric current doesn't heat the wire evenly. Connections to the nitinol will

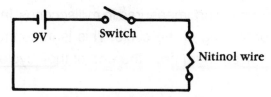

Figure 11-2. A schematic for direct electric heating.

draw heat away from the ends of the wire. This results in the center of the wire heating faster than the ends. So although direct electric heating works, a better method is pulse width modulation.

PULSE WIDTH MODULATION HEATING

Heating the wire is more efficiently controlled with pulse width modulation (PWM) heating. Here we use a square wave from a simple circuit to turn on and off the electric current. Depending upon the frequency and duty cycle of the square wave, we can adjust the amount of contraction and maintain the wire in a contracted condition for a longer period of time. Because of the rapid on and off the wire has time to distribute the heat and results in a more uniform heating. This is the method that we shall use.

CIRCUIT

Usually a 555 timer is used to provide a square wave to activate nitinol wire. Although this is a good standalone method, it wouldn't allow for easy interfacing to a computer. The circuit we shall use is designed around a 4011 Quad NAND gate, see **Figure 11-3**. The NAND gate is made to generate a square wave that can be operated as a standalone circuit using a switch, or it can be interfaced to any computer by connecting the circuit to a port line that can be brought high and low under program control. The output of the 4011 is connected to an NPN transistor which is capable of switching a higher current than is required of the nitinol wire.

The circuit can be wired on a prototyping breadboard, using a manual switch to activate the nitinol.

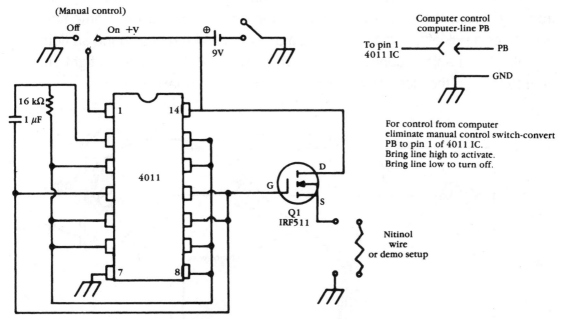

Figure 11-3. A schematic for a pulse width modulation circuit.

NITINOL DEMONSTRATION

To demonstrate the potential of this material we need to build a small mechanical device. If you're like me you'll want the simplest unit to start with. To make our electric muscle, the materials you'll need are three machine screws with six nuts, a piece of perf board or plastic, a small rubber band and, of course, a length of nitinol material.

The machine screws, nuts and perf board are available from Radio Shack, see parts list. The nitinol wire is available from Images Company, see suppliers index in Appendix A.

Look at **Figure 11-4**. Drill three holes in the perf board to accommodate the machine screws in a triangle pattern as shown. The nitinol wire is connected to the two top screws. The rubber band is looped around the bottom machine screw, with the nitinol wire looped through the top of the rubber band. To determine the distance you should place the bottom machine screw, stretch the rubber band from a position that is parallel with the top screws and down. Remember the nitinol has a pull of about 11 ounces so don't make the rubber band so tight that the nitinol isn't able to contract and move upwards. But it should be tight enough for it to take up the slack of the nitinol wire when it is relaxed.

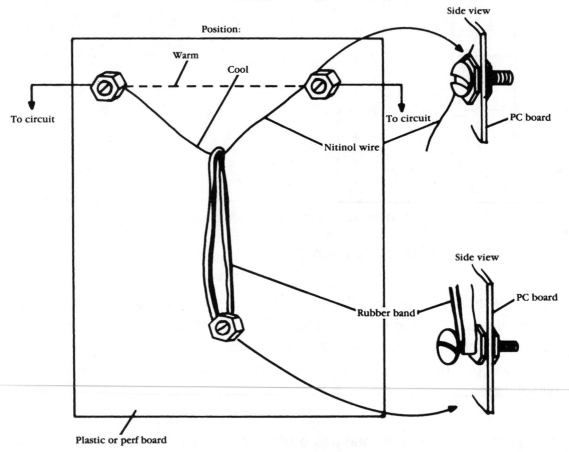

Figure 11-4. The test fixture.

For the connections from the circuit to the nitinol I used small jumper cables connected to the back of the machine screws to the circuit. You could of course simply use wire.

When the unit is activated the wire gets hot, contracts and pulls up from the rubber band. When the unit is deactivated the wire cools, elongates and lowers into its resting position.

USE

When you have the circuit wired and the demo electric muscle unit built; apply power to the circuit. The control switch allows you to contract the muscle by putting the switch in the + Voltage setting. Connecting the switch to ground will turn off the square wave generator and the electric muscle will relax. My unit performed slowly, probably due to the high tension I had on the wire. I also had no hesitation about overdriving the wire; I'm not concerned with making the wire last for one million cycles. You still should be careful not to overheat the wire. You can check if you're overheating the circuit by touching the transistor; if it is hot you can assume the wire is overheating. To reduce the current add another 10 ohm resistor in line with the first. This will reduce the current.

2ND DEMONSTRATION MODEL

The second nitinol actuator shows how you can amplify the mechanical motion of the wire using a lever, see **Figure 11-5**. The lever pivots on screw A. The nitinol wire is attached to the lever and screw E. The wire is threaded around the other screws B, C and D. When activated, the lever rises. The lever can be made from any material; wood, plastic or metal. If you use a metal lever you can make electrical connection to the wire through screws A and E. If the lever is made from a non-conductive material, use screws B and E for the electrical connections. If the lever is too light you may need to add some weight on the end of it so that it lowers easily when the nitinol is deactivated.

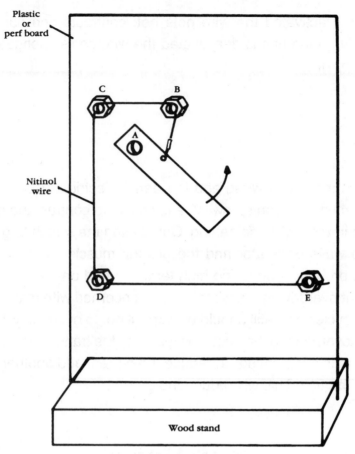

Figure 11-5. The lever.

GOING FURTHER

We have just scratched the surface of the applications that can be derived using this material. It is quite possible to build a realistic android hand. As an example a simple digit flexor is illustrated in **Figure 11-6**. This unit is constructed of a three-hole soft rubber or silicone tubing. The nitinol wire is threaded in a loop through the two outer holes. A copper wire is threaded up through the center hole. The loop of nitinol wire and the end of the copper wire is crimped in a small terminal, see **Figure 11-1**. By applying current between the copper wire and one end of the nitinol you can make the tube flex right (A-C) or left (B-C). Or by applying power to the two ends of the nitinol wire the tube will flex backwards.

Heat engines are another fertile field for experimentation and development. One company sells a toy boat powered by nitinol wire. The boat has a small cargo bay for ice. The difference in temperature between the ice and the water the boat rests in powers the toy boat.

You may want to attempt to train a piece of nitinol wire to a particular shape. You can do this by bending the wire to the shape you want, clamping it in position and heating it to about 540°C. You also might want to try direct electric heating of the wire to reach annealing temperature.

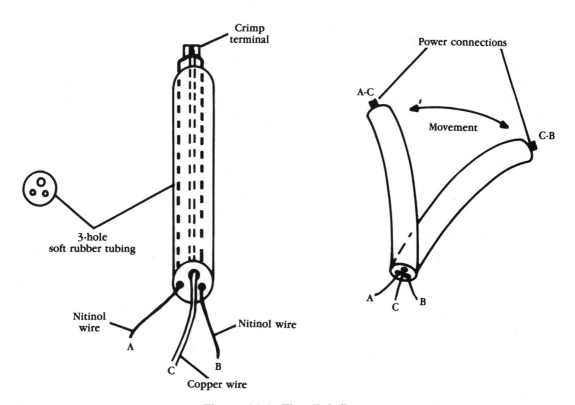

Figure 11-6. The digit flexer.

PARTS LIST

◆

ITEM		SOURCE
Nitinol Wire 12"		Images Company
PWM CIRCUIT:		
4011 Quad NAND Gate	#276-2411	Radio Shack
1 mF Capacitor	#272-1434	Radio Shack
15K Resistor	#271-036	Radio Shack
2N2222 Transistor (NPN)	#276-1617	Radio Shack
Switches, 9V Battery & Cap		
DEMO 1 & 2:		
Round Head Machine Screws 6-32 x 3/4"	#64-3012	Radio Shack
Hex Nuts 6-32	#64-3019	Radio Shack
Perf Board (or plastic)	#276-147	Radio Shack

◆

CHAPTER

---◆---

AIR
POLLUTION MONITOR

In this project we will build an air pollution monitor that can sniff and detect various airborne compounds. The heart of the circuit is a TGS gas sensor. The circuit can easily be interfaced to a computer and would allow the computer to sense its environment for a number of toxic compounds.

The most obvious use for the simple circuit is as an air pollution monitor or automatic ventilation control.

SEMICONDUCTOR SENSOR

The sensor material is an N type sintered SnO_2 (Tin Dioxide). When a combustible or reducing gas is absorbed on the sensor's surface, the resistance of sensor decreases dramatically. This makes it an easy job to build a detection circuit or computer interface.

The SnO_2 must be heated to a high temperature for it to become useful as a gas sensor. The sensor contains a small internal heating coil that heats the SnO_2 material inside the sensor between 200-400°C.

There are many other sensor materials that perform just as well as SnO_2, but the SnO_2 is the preferred material of use because of its chemical stability.

GASES DETECTED BY SENSOR

Inorganic Gases: Ammonia, carbon monoxide, hydrogen, hydrogen cyanide.

Hydrocarbons & Derivatives: Methane, ethane, propane, butane, pentane, hexane, heptane, octane, decane, petroleum ether, petroleum benzine, gasoline, kerosene, petroleum naphtha, acetylene, ethylene, propylene, butadiene, butylene, benzene, toluene, o-xylene, m-xylene, ethylene oxide.

Alcohols: methanol, ethanol, n-propanol, isopropanol, n-butanol, isobutanol.

Ethers: Methyl ether, ethyl ether.

Ketones: acetone, methyl ethyl ketone.

Esters: Methyl acetate, ethyl acetate, n-propyl acetate, iso-propyl acetate, n-butyl acetate, isobutyl acetate.

Nitrogen Compounds: Nitro methane, mono methyl amine, dimethylamine, trimethyl amine, mono ethyl amine, diethyl amine.

◆

Halogenized Hydrocarbons: Methyl Chloride, methylene chloride, ethyl chloride, ethylene chlorhydrin, ethylidene chloride, trichloro ethane, vinylidene chloride, trichloro ethylene, methyl bromide, vinyl chloride.

Although the sensor is capable of detecting these compounds it can not tell us which gas of the many it is detecting. There is ongoing research and development on constructing sensors that can determine particular compounds.

FUTURE SENSORS

Humans do not have a acute sense of smell when compared to other mammals such as dogs. Even so, humans have about 2 million olfactory sensors in the nasal cavity that can detect thousands of different smells.

It's interesting to note that the nose and nasal passages originally had nothing to do with respiration, but evolved separately as a smelling device. Smelling is a primitive function, and our smelling apparatus evolved to respiratory passages.

For instance, fish have nostrils, but they do not breath through them. To breathe, fish take water in through their mouth. The nostrils are used for smelling (sensing) and locating food.

It may be some time before artificial sensors meet the diversity and sensitivity of even the human olfactory. But what artificial sensors lack in diversity they may more than make up for in sensitivity. One scheme under development uses 12 SnO_2 sensors simultaneously. Each of the sensors is slightly different from the others. Naturally, each sensor reacts a little differently when detecting a particular compound. This information is gathered simultaneously from the twelve sensors and is stored as a gas profile in a computer's memory bank. When an unknown gas is presented for analysis, not only is the gas detected, but the type of gas is determined by matching its unique profile from the profiles stored in the computer's memory.

The uses for these more sophisticated sensors are varied. They will be used in food processing equipment to check for food spoilage.

The fragrance industry can use them to mix and match fragrances or to make inexpensive perfumes that more closely match expensive perfumes.

Biologists may use them to detect and measure human pheromones or hormonal changes in humans and other mammals.

SENSOR CHARACTERISTICS

Figure 12-1 is a cutaway illustration of the sensor. The sensor has the appearance of a short 6-pin orange plastic tube. The top and bottom of the sensor is covered with 100 mesh stainless steel wire cloth. The heart of the sensor is the cylindrical form in the middle of the unit. The cylinder is a ceramic material with the SnO_2 material deposited on its surface. The heater coil is located inside the ceramic cylinder. The heater has a resistance of 38 ohms.

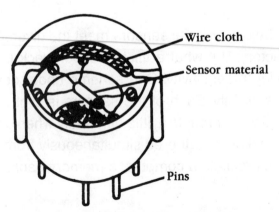

Wire cloth

Sensor material

Pins

Figure 12-1. A cutaway view of the gas sensor.

Figure 12-2 shows the pin-out schematic of the sensor. Pins 1 and 3 are internally connected as are pins 4 and 6. Pins 2 and 5 connect to the heater coil. The heater is not polarized so the + 5 supply voltage for the heater may be connected to either pin.

Figure 12-3 shows the bottom of the sensor and pin location.

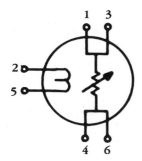

Figure 12-2. A schematic of the electrical-equivalent gas sensor.

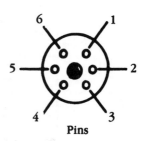

Figure 12-3. A bottom view of the pin-out gas sensor.

The heater requires a regulated +5 volts for proper operation and draws 130 mA. This regulated voltage is easily supplied using a 7805 regulator. The heater coil heats the SnO_2 material. The circuit voltage passing through the SnO_2 sensor material should not exceed 24 volts. The specification on the sensor states that the sensor may be used with AC or DC voltage. The circuits described here use only DC power.

When the sensor is first used, it requires a 15 minute break-in period. The break-in period is necessary whenever the sensor hasn't been used for some time, as will be the case with the one you receive.

TEST CIRCUIT

Figure 12-4 shows a simple circuit. The reading on the voltmeter in this setup is about .5-1 volt (after break-in period). When the sensor detects any compound, its resistance drops causing the reading on the voltmeter to jump up to 5 volts or more. The actual meter reading depends upon the concentration of the gas.

Use the circuit to test the response to various compounds you have around your home, such as: glue, rubber cement, bleach, alcohol, cleaning fluid, etc. It's enough to bring the material close to the sensor for a reaction; do not spray anything on it.

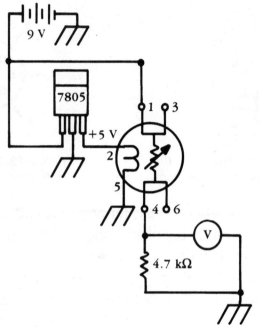

Figure 12-4. A schematic of the test circuit.

SIMPLE AIR POLLUTION MONITOR

Figure 12-5 is a simple go/no-go air pollution monitor. The circuit uses two comparators off a standard 339 Quad comparator IC. The green LED is lit as long as the gas sensor doesn't detect any airborne compounds. When the circuit does detect a gas, the green LED turns off and the red LED turns on.

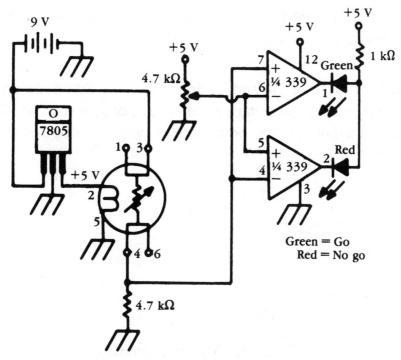

Figure 12-5. A schematic of the air pollution monitor.

There is nothing critical about the circuit, point-to-point wiring is fine. Although you can solder wires directly to the TGS gas sensor pins, it may be a better idea to purchase a socket along with the sensor. The socket makes it easy to mount the sensor on the outside of the small instrument case.

When the circuit is completed it needs to be calibrated. This is a simple procedure. Turn on the circuit, allow a warm-up period. If this is the first time the sensor is used wait 15 minutes before you calibrate the circuit. If not, 2-5 minutes will suffice. To calibrate, adjust potentiometer R2 to the point where the green LED is about to go off and the red LED is about to turn on. That's it; it's calibrated. Pretty simple, huh?

To test, bring the sensor close to your mouth and breath on it. The green LED should go out and the red LED on. The sensor is detecting the CO_2 from your breath. The LEDs should go back to their normal mode in 15-30 seconds depending upon air circulation.

Another test you may want to try is releasing some gas from a butane lighter (unlit), or opening a bottle of alcohol by the sensor.

Since the heater section of the sensor draws 130 mA, I'd advise using an 9V alkaline battery for portable operation. However, any battery you use will wear out pretty fast with constant use. The best power supply is a plug-in transformer to supply 9 VDC to the circuit.

GOING FURTHER

The sensor can be used for:
       Gas Leak Detector
       Carbon Monoxide Detector
       Automatic Ventilation Controller
       Fire Alarm (detects combustible gases in smoke)
       Alcohol Detector
       Air Pollution Monitor

This sensor can easily be interfaced to a computer. The comparator is set to trip when the resistance of the sensor drops. (This is the equivalent of the red LED section of our simple air pollution monitor). The comparator will output a +5 V signal which can be read off one of the computer's interface lines. When the sensor detects a gas, the +5V line will drop to ground.

By connecting an analog to digital converter in place of the comparator, the computer can read the changes in resistance and thus the concentrations of the gas it is detecting.

Sensitivity can be improved by placing the sensor in the arm of a balanced resistance bridge. The output of the bridge is then connected to a standard op-amp for amplification. When the sensor detects a minute amount of gas, the bridge is thrown out of balance, and the resulting signal is amplified by the op-amp.

◆

PARTS LIST

	ITEM		SOURCE
R1	4.7K	#271-1330	Radio Shack
R2	4.7K Potentiometer	#271-281	Radio Shack
R3	1K	#271-1321	Radio Shack
IC1	7805 Voltage Regulator	#276-1770	Radio Shack
IC2	339 Comparator	#276-1712	Radio Shack
D1	Green LED	#276-022	Radio Shack
D2	Red LED	#276-041	Radio Shack
	Snap in Holders for LEDs	#276-079	Radio Shack
822	TGS Gas Sensor/Data Sheet		Images Company
822A	Socket for TGS Sensor		Images Company
Misc	Instrument Case, Switch, PC Board, 9V Battery, Cap		

CHAPTER

---◆---

NEURAL
NETWORKS

Neural networks are computer operating systems that function and learn based on the biological models of the brain. The first question that comes to mind is why model the brain. Computers are functioning pretty well without neurons. Well, computers are quite limited when asked to do some simple things in real time, like identifying speech or a picture. The ability of neural networks to accomplished these (and other) tasks in real time is best described by an analogy.

WHY NEURAL NETS

Suppose your friend is standing about twenty feet away from you. He looks at you and yells "catch!" as he throws a baseball. You see the ball coming, move to the right, raise your arm, the ball hits your hand and you catch it.

Although this is a simple task for humans, to program the actions in a computer is extremely hard. The factors needed to be programmed are: 1) estimating the initial velocity of the ball; 2) calculating the trajectory through three-dimensional space; 3) moving a catching unit from a rest point in three-dimensional space, to another point in three-dimensional space, where the ball will hit, and; 4) doing it in real time to catch the ball. If we asked a computer to do all this, the ball would be rolling twenty feet behind it as it was calculating the initial velocity and trajectory of the ball.

Neural nets, on the other hand, do not perform calculations. In this way they act like a biological system. If it were your job to catch the ball, you would not be standing there performing calculus equations to see where the ball would end up. You would instinctively, without any math whatsoever, estimate where the ball is going and catch it. How did you accomplish this task without using math. You learned it. Probably by missing hundreds of balls thrown to you, but remembering what you did when you actually caught or came close to catching the ball.

It's the same way with neural nets; they are taught. To teach a neural net to catch a ball, we don't fill it up with equations. We assign it a task - "catch the ball" - then we keep throwing balls to it. Eventually, even if its motions are completely random in the beginning, every time it comes close, even by accident, that procedure is reinforced (strengthened) in the neural network (supervised learning). The more a certain procedure is reinforced the more likely it will occur. With continual training the robot will always come close to catching the ball and occasionally hitting the catching mit. Soon the ball will always hit the catching mit. And finally the neural network will learn how to catch. All without a single equation.

BIOLOGICAL NEURONS

The brain is a pretty sophisticated piece of wetware. It is made up of special cells called neurons (nerve cells). The human brain has about 10^{12} (1,000 billion) neurons. Neurons have inputs called dendrites and outputs called axons, see **Figure 13-1**. The axons connect to the dendrites of other neurons. There isn't a precise number of inputs or outputs to any neuron. A neuron may have a thousand inputs and a single output, vice versa or anything in between. For instance, a brain cell may have so many dendrites coming out of the cell that it looks like tree. Motor neurons have long axons running from the central nervous system to the muscle. Sensory neurons generally have a single fiber.

The connection between the dendrite of one neuron and the axon of another is called a synapse. This is where information is transmitted. The transmission speed is slow by computer standards, but the brain has a tremendous advantage over computers, parallelism. Computers perform operations serially, one after another after another, in the standard Von Neumann computer architecture. The brain, however, operates with massive parallel structures.

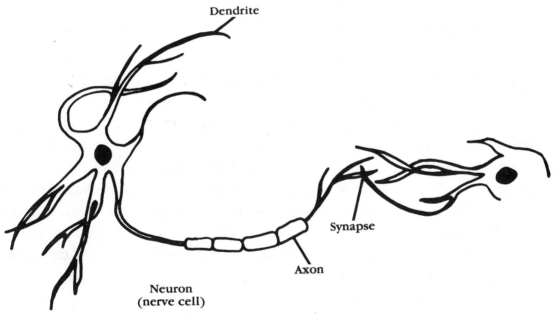

Figure 13-1. The biological neuron.

HUMAN BRAIN

The human cerebral cortex has about 100 billion (10^{11}) neurons. Each cell has about 1,000 dendrites that make up about 100,000 billion (10^{14}) synapses. If we assume the brain operates at about 10 Hz, it performs some 1,000,000 billion (10^{15}) interconnects per second, to which we humbly add that the brain weighs in at about 3 lbs.

As you can imagine, to build a neural structure that mimics the brain is beyond our current capabilities.

These massive parallel structures allow us to do things that are very hard or impossible to for a computer to do. For instance: think, see, identify objects, faces, colors, hear, remember, act and create. Simple language processing alone, far exceeds any computer capabilities to date.

IN THE BEGINNING

The neurobiologists who study the brain made mathematical models of nerve cell behavior. Based on these mathematical models work began in the 1940's and 1950's to build computer devices that model some aspects of the human nervous system. Reasonable success had been achieved by Frank Rosenblatt with his Perceptron work. However, at that time in computer history, there was a division in thought as to what avenue of computer research would lead the way to developing artificial intelligence in computers, neural networks or rule based expert systems. In the ensuing battle for research funding, neural net research was pretty much halted in the early 1960's by a critical paper from the then-current computer experts Minsky and Papert.

Neural net research didn't pick up again until 1982, when John Hopfield showed that the XOR limitations reported in Minsky and Papert paper were only true for the most primitive two-layer neural networks. The fact that the rule based expert systems promoted by Minsky and Papert did not live up to their proposed expectations and were nowhere close, despite 20 years of exclusive funding, research and development, helped revive interest in neural networks.

MODELS

There are many models and various learning systems using neural networks. **Figure 13-2** illustrates a simple neural net that separates different fruits and vegetables.

We assume the input neurons to the computer can determine basic shape and color. The lines between the inputs and outputs represent connections whose weights can be adjusted. Although it isn't shown in the diagram, there are connections between each input neuron and middle layer neuron (fully connected). However, just the strong connections are shown with lines.

To start teaching, the first pattern is introduced to the inputs - let's say red and round, the target answer is apple. The weights or interconnections between the input/output are modified so that these two properties converge to apple. This is shown by the lines between red and round inputs to the middle layer apple. All other connections are made weaker and do not show on the diagram. Similarly, other patterns are introduced and the weights adjusted to output the correct answer.

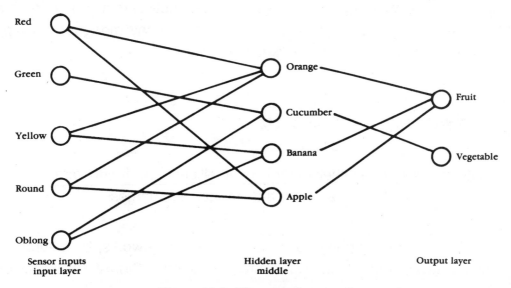

Figure 13-2. The neural network.

COMPUTER MODELING

Neural computers are still in their infancy. Manufacturers and chip designers are beginning to introduce neural and fuzzy logic chips. So far, most neural networks are modeled on standard serial computers.

It is important to note that it is doubtful that neural networks will be used exclusively in designing computers and programs. Future computers will use procedures from the rule based expert system databases as well as parallel neural networks using a collaboration of techniques to perform functions, choosing and using whatever technique is most appropriate for the task at hand.

As an example, you would not use a neural network to calculate your tax return. The fuzzy logic would be imprecise. In this case you would used a standard serial rule based program. If you're trying to identify a rock, using measured properties, like hardness and density, you may end up using an expert system. Neural networks are more appropriately used for other tasks like voice recognition and pattern identification.

There are some computational tasks where neural networks are superior to serial rule based programs, and these are being utilized today: mortgage loan applications, life insurance and stock market analysis.

ELECTRONIC NEURONS

Individual neurons by themselves are not intelligent, but if you wire billions of them together as in the human brain an intelligence emerges.

We will start at the most primitive concept in neural networks, the neuron. The electronic equivalent of a biological neuron is shown in **Figure 13-3**. There are different types of neurons. Neurons can perform summation, difference or signal inversion of the inputs. How the neuron responds to the input is based upon its threshold value. When the threshold value is met or exceeded it becomes active. Activated neurons may also respond differently. Some neurons are excitatory meaning that they fire when stimulated, others are inhibitory, or don't fire when stimulated. Some neurons are stronger than others; to use the proper terminology they "carry more weight", and are able to stimulate or inhibit more neurons than others. Typically, neurons are not digital in nature, although they can be simulated with digital circuits.

We will build a simple electronic neuron and use it as a self-contained control system.

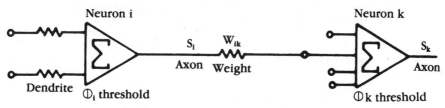

Figure 13-3. An electronic neuron.

SUN TRACKER NEURAL CIRCUIT

The circuit illustrated in **Figure 13-4** is a basic neural net. The purpose of this net is to steer or look toward a light source like the sun.

The operation of the circuit is simple. Two cadmium sulfide photoresistors act as neural sensors forming neural inputs to our neuron, which is the 741 op-amp. **Figure 13-5** illustrates how the input operates. As long as the sun is directly aligned with the two photoresistors they are equally exposed, and the inputs to the neuron balance out. As the sun moves across the sky the alignment is thrown off, making one of the inputs stronger than the other. The 741 op-amp neuron activates a small DC motor, moving the tracker up or down, depending upon which input is stronger, to bring the tracker back into alignment.

◆

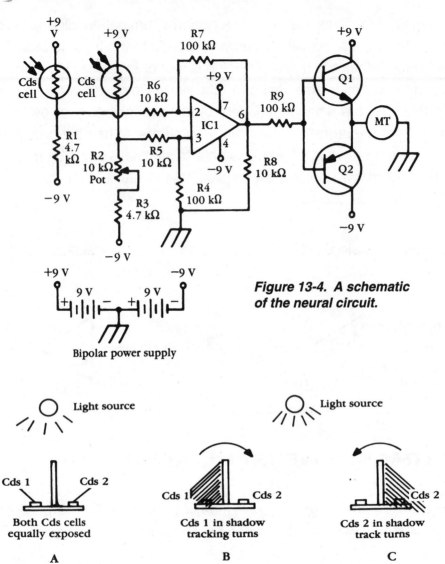

Figure 13-4. A schematic of the neural circuit.

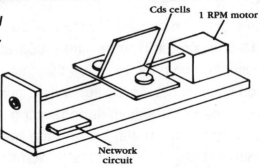

Figure 13-5. A detail of the neural input (A, B, C) and the prototype (D).

A — Both Cds cells equally exposed

B — Cds 1 in shadow tracking turns

C — Cds 2 in shadow track turns

D

The two transistors Q1 and Q2 that read the signal from the 741 op-amp neuron and activate the motor may also be looked upon as neurons, becoming active only when their threshold is reached.

The motor used in the prototype is a 1 rpm 12 volt DC motor. Just about any low voltage DC motor can be used. Radio Shack sells inexpensive hobby motors, however the rpm on these motors are pretty high. If you use one of these motor you will need to gear it down somewhat, see **Figure 13-6**.

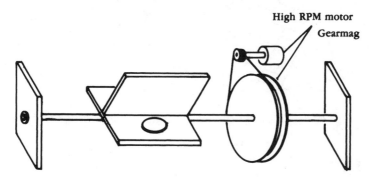

Figure 13-6. The gearing for the high rpm motor.

If this circuit is used with an artificial light source no modifications are necessary. If you use it to track the sun you may have to cover the photocells with a colored plastic to cut down on the light intensity. The sun is such a strong light source that it will easily saturate the photocells.

To train the circuit, expose both photocells to equal light, adjusting potentiometer R2 until motor stops. To test, cover one photocell and the motor should begin rotating. Uncover it and it should stop. Then cover the other photocell and the motor should begin rotating in the opposite direction. At this point connect the motor to the unit. If the motor is turning in the opposite direction you need to keep the tracker aligned just reverse the power wires to the motor.

The circuit has immediate practical applications in the field of solar energy. If we wish to track the sun to obtain the maximum output from solar cells, a furnace, water heater or any solar gathering device, the sun tracker will faithfully follow the sun, see **Figure 13-7**. Notice this experimental neural circuit tracks a light source without using any equations.

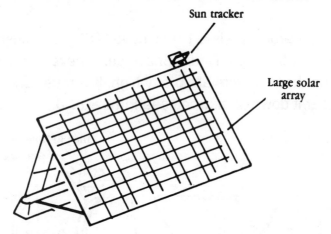

Figure 13-7. Using the tracker with a large solar array.

OTHER USES

The principal of this simple neural net can be applied to other tracking problems.

As the network stands it can move up or down to track a light source. A similar net can be incorporated that operates in the same manner for horizontal directions, see **Figure 13-8**. If we changed the four inputs from photocells to radio antennas, we could track radio-emitting satellites across the sky. Since the net is self-correcting, once it's locked onto the satellite, we do not need to know its orbit.

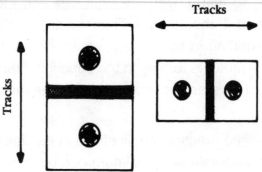

Figure 13-8. The configuration for using four neural inputs.

Obviously this type of self-correcting tracking system has military potential. Change the inputs to small independent radar systems and you have a ground tracking device for incoming aircraft, missiles and rockets. The ground tracking could be used for aiming defense systems such as missiles or lasers.

Place a miniature neural system on a missile connected to its flight control and you have ground-to-air or air-to-air on-the-fly tracking. Using the same with sonar systems you can create smart torpedoes.

This is basic stuff. Adding a little more AI (artificial intelligence) to the system for object identification, you can prevent downing your own aircraft.

PARTS LIST

	ITEM		SOURCE
IC1	741 Op-Amp	#276-007	Radio Shack
R1, R3	4.7K	#271-1330	Radio Shack
R2	10K Potentiometer	#271-282	Radio Shack
R4, R5, R8	10K	#271-1335	Radio Shack
R6, R7	100K	#271-1347	Radio Shack
R9	100 ohm	#271-1311	Radio Shack
Cds	Photocells (5/pak)	#276-1657	Radio Shack
Q1	NPN Transistor	#276-1617	Radio Shack
Q2	PNP Transistor	#276-1604	Radio Shack
MT	Motor, see text		

CHAPTER

---◆---

INTRODUCTION TO STEPPER MOTORS

Stepper motors are commonly used in robotics, automation and positioning control in commercial and industrial equipment. If you own a computer these motors are as close to you as your disk drive and printer. Stepper motors are used in these applications because they are easily controlled by digital circuits, and most importantly, capable of precise positioning. We will build a simple stepper motor interface to examine the basic operating principles of stepper motors.

Stepper motors are different from the normal electric motors. When you apply power to an ordinary motor the rotor turns smoothly. A stepper motor, however, runs on a sequence of electric pulses to the windings (or phases) of the motor. Each pulse to the winding turns the rotor by a precise amount. These pulses to the motor are often called steps. Stepper motors are manufactured with different amounts of rotation per step (or pulse), depending upon the application it is designed and built for. The specifications of the stepper motor will state the degree of rotation per step. The range of rotation per step can vary from a fraction of a degree (i.e. .72 degree) to many degrees (i.e. 22.5 degrees).

BASIC OPERATIONS

Figure 14-1 is of a stepper motor stepping through one rotation. Stepper motors are constructed of strong permanent magnets and electromagnets. The permanent magnets are located on the rotating shaft, called the rotor. The electromagnets or windings are located on the stationary portion of the motor, called the stator. The stator surrounds the rotor.

In **Figure 14-1** position I we start with the rotor facing the upper electromagnet that is on. Moving in a CW (clockwise) rotation the upper electromagnet is switched off as the electromagnet on the right is switched on. This moves the rotor 90 degrees in a CW rotation, shown in position II. Continuing in the same manner, the rotor is stepped through a full rotation till we end up in the same position as we started, shown in position V.

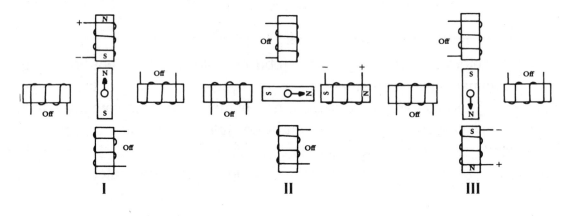

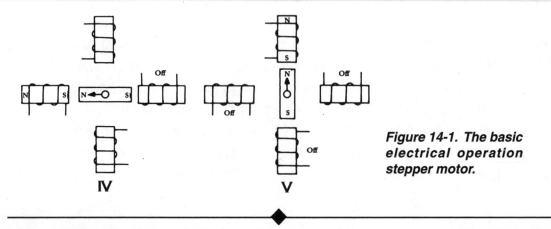

Figure 14-1. The basic electrical operation stepper motor.

RESOLUTION

The degree of rotation per pulse determines the resolution of the stepper motor. In the illustrated example the rotor turned 90 degrees per pulse, not very practical. A real world stepper motor, for instance one that steps 1 degree per pulse, would require 360 pulses to achieve one revolution. Another motor with less resolution (greater degree per step), that steps, say, 3.75 degrees per pulse, would only require 96 pulses for one full rotation.

Without getting into gearing or gear ratios, let's assume that the stepper motor is used for positioning in a linear motion table, and further that each revolution of the motor is equal to one inch of linear travel on the table. It becomes apparent that each step of the motor defines a precise increment of movement.

In making a comparison between the two stepper motors, the ability to locate and position more precisely on the table would be with the stepper motor with the higher resolution (one that requires the most steps per revolution). For the motor that steps 3.75 degrees per step, the increment of movement is approximately .01 inch per step. If this resolution is sufficient for your table it's fine to use this stepper motor. If, however, you required greater resolution, the 1 degree per step motor would give approximately .0027 inch per step. So you see the increment of movement is in proportion to the degrees per step.

HALF STEP

It is possible to double the resolution of some stepper motors by half stepping. In **Figure 14-2** this is illustrated. In position I, the motor starts with the upper electromagnet switched on, as before. In position II the electromagnet to the right is switched on while keeping power to the upper coil on. Since both coils are on, the rotor is equally attracted to both electromagnets and positions itself in between both positions (a half step). In position III the upper electromagnet is switched off and the rotor completes one step. Although I am only showing one half step, the motor can be half stepped through the entire rotation.

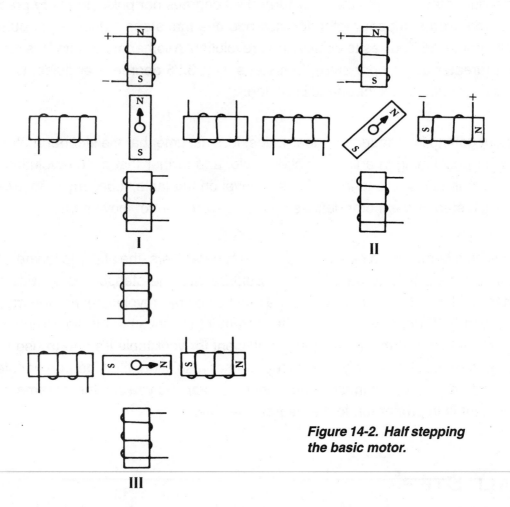

Figure 14-2. Half stepping the basic motor.

OTHER TYPES OF STEPPER MOTORS

You may run across a 4-wire stepper motor. These stepper motors have two coils, with a pair of leads to each coil. Although the circuitry of this stepper motor is simpler than the one we are using, it requires a more complex driving circuit. The circuit must be able to reverse the current flow in the coils after it steps.

REAL WORLD

As stated, the stepper motors diagrammed for illustration wouldn't be of much use in the real world, rotating 90 degrees per step. Real world stepper motors employ a series of mini-poles on the stator and rotor which improve the resolution of the stepper motor. Although **Figure 14-3** may appear more complex than the previous figures, it is not really so. Its operation is identical, and to prove the point we'll step through the illustration.

In **Figure 14-3** the rotor is turning in a counterclockwise (CCW) rotation. In position I the north pole of the permanent magnet on the rotor is aligned with the south pole of the electromagnet on the stator. Notice that there are multiple positions that are all lined up. In position II the electromagnet is switched off and the coil to its immediate left is switched on. This causes the rotor to rotate CCW by a precise amount. It continues in this same manner for all the steps. Examine the pole relationship between IV and V. The rotor is still moving CCW, and in position V the stator poles are in the same orientation as position I. This is where the sequence of electric pulse would begin to repeat themselves and keep the rotor turning CCW.

Figure 14-4 illustrates the half step with the multi-pole position. It is identical to the half step described before.

TEST CIRCUIT

Although we are going to connect the stepper motor to a dedicated IC, I feel it would be a good idea to build a manual controller first. With this manual circuit, you can check and verify your wiring of the stepper motor before building the Amiga interface. In addition, it's an excellent tool to use if you're checking out a different stepper motor from the one used in this article.

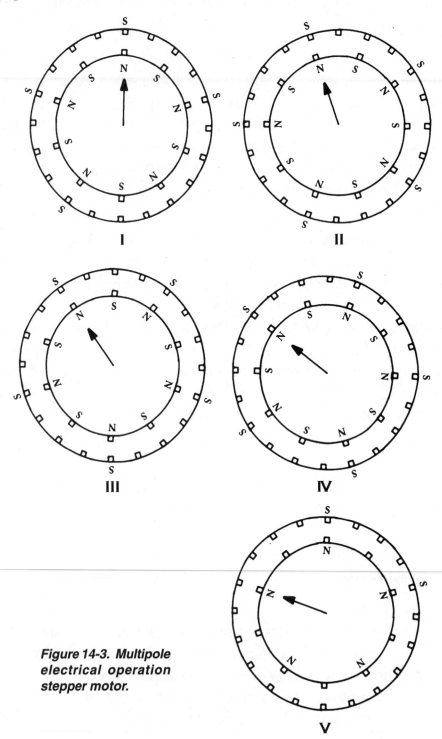

Figure 14-3. Multipole
electrical operation
stepper motor.

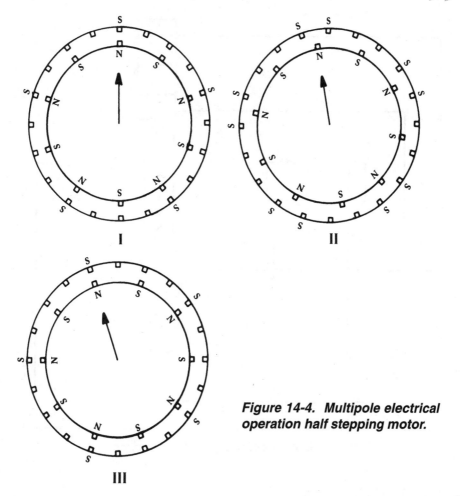

I

II

III

Figure 14-4. Multipole electrical operation half stepping motor.

Look at **Figure 14-5**; the circuit is an epitome of simplicity. Switches S1 thru S4 are normally open subminiature push button switches, see parts list. The four switches allow you to drive the stepper motor manually. By changing the sequence of the steps we can do full step and half step increments in either direction (CW or CCW). The diodes D1 thru D4 are used to prevent sparking and protect the balance of the circuit. These diodes become more important when the motor is interfaced to a computer. The batteries used in the circuit are two small 12V batteries in series to generate the 24 volts required. The stepper motor we are playing with is a 24 volt model. The rectangular box at the top of the diagram with six screws is six PC board terminals interlocked together, see parts list. This simplifies connecting the wiring from the motor to the circuit.

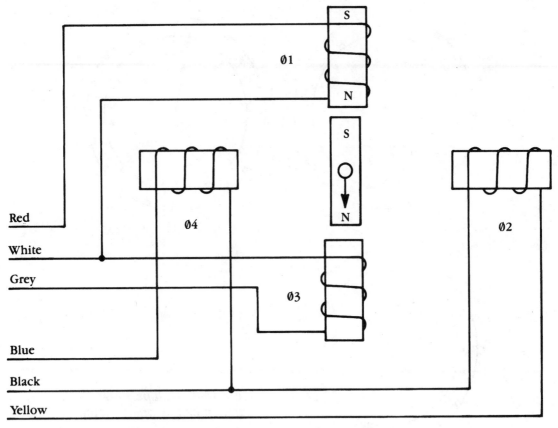

Figure 14-5. An electric equivalent stepper motor.

STEPPER MOTOR

Figure 14-6 is an electric equivalent circuit of the stepper motor we are using. The stepper motor has 6 wires coming out from the casing (see **Figure 14-7**). Three leads go to each half of the coil windings, and that the coil windings are connected in pairs. If you just picked this stepper motor and didn't know anything about it, the simplest way to analyze would be to check the electrical resistance between the leads. By making a table of the resistances measured between the leads you'd quickly find which wires are connected to which coils.

On the motor we are using there is a 60 ohm resistance between the center tap wire and each end lead, and 120 ohms between the two end leads. A wire from each of the separate coils will show an infinitely high resistance (no connection) between them. Armed with this information you can tackle just about any 6 wire stepper motor you come across. The stepper motor we are using rotates 3.75 degrees per step.

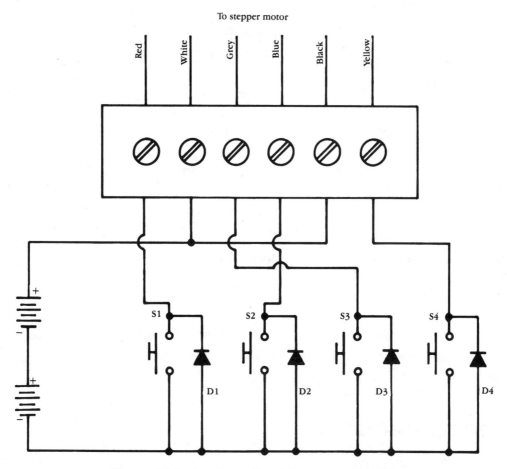

Figure 14-6. A schematic of the stepper motor.

Figure 14-7. The stepper motor circuit.

TEST CIRCUIT DEMO

After you are finished wiring the test circuit and connecting the stepper motor, use the following two tables to step or half step the motor.

FULL STEP			
S1	S2	S3	S4
on	-	-	-
-	on	-	-
-	-	on	-
-	-	-	on

HALF STEP			
S1	S2	S3	S4
on	-	-	-
on	on	-	-
-	on	-	-
-	on	on	-
-	-	on	-
-	-	on	on
-	-	-	on
on	-	-	on

When you reach the end of the table the sequence repeats starting back at the top of the table.

If you want to reverse the direction of travel, just reverse the sequence of the table, starting from the bottom and working to the top.

TROUBLESHOOTING

If you use the stepper motor listed in the parts list I don't think you'll run across any problems. If you should, the first thing to check is the diodes. Make sure you have them in properly, facing in the direction shown in the schematic.

If the stepper motor moves slightly and quivers back and forth, chances are the batteries are too weak to power the motor, so replace them with fresh batteries. The batteries do wear out pretty quickly. This problem will be solved if you modify your existing circuit board to use line voltage with a step-down transformer.

GOING FURTHER

There are a number of dedicated integrated circuits available for powering and controlling stepper motors. Next time we will use a dedicated IC to make a practical device for astrophotography.

PARTS
LIST

◆

	ITEM		SOURCE
IM-Step	Stepper Motor 6-wire	$20.00	Images Company
D1-D4	1N914 Diode (50/pak)	#276-1620	Radio Shack
PC Board	Terminals Stackable	#276-1388 and #276-170	Radio Shack
S1-S4	Momentary Push Switch (4/pak)	#275-1547c	Radio Shack
	Battery Holder	#270-405	Radio Shack
	Battery	#23-144	Radio Shack

◆

CHAPTER

---◆---

EQUATORIAL CAMERA MOUNT FOR ASTROPHOTOGRAPHY

We have all seen them, strikingly beautiful photographs of the night sky. Stars more abundant than you have ever seen, some photos showing brilliantly colored gaseous nebula.

The magnificence of these photos may drive you outside in the middle of the night to attempt to capture these spacescapes for yourself. You grab a tripod and your trusty 35mm camera and you're off. What happens after the film is developed can be, to say the least, disappointing. What happened to those great photographs of the night sky?

If you anticipated that extra time is needed for the faint light of the stars to expose the film, you're right. But if you kept the camera shutter open for more than 10 seconds, the stars started to make circular trails on the photograph due to their apparent motion. This time restraint severely limits the quality of photographs you may take. Shorter exposures will eliminate the star trails, but the resulting photos will not show any nebula, and the stars will not be numerous or bright - rather unimpressive photographs.

The simple equatorial mount described here will allow you to improve you night sky picture taking. With it you can make exposures as long as 30-40 minutes without any star trails. These longer exposures will allow you to capture details in the night sky you ordinarily cannot see.

The principal of the equatorial mount is simple. It moves the camera at the same rate as the stars appear to move across the sky. The apparent motion of the stars is caused by the rotation of the earth. This apparent motion of the stars across the sky is approximately 15.041 degrees per hour. The motion of the stars will appear greatest at the horizon and least near the pole star Polaris, (Northern Hemisphere).

To explain this better, imagine the earth stationary, with the stars in the sky printed on a transparent sphere surrounding the earth, see **Figure 15-1**. Now picture this transparent sphere rotating counterclockwise on a pivot point located at the pole star Polaris. You now have a pretty good mental image of how the stars move across the sky.

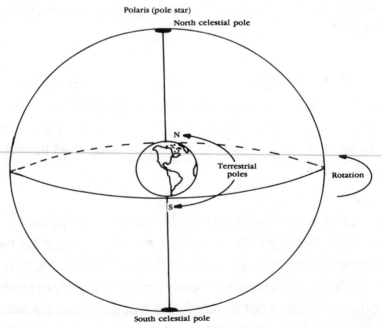

Figure 15-1. A celestial sphere which shows Earth's rotation.

The type of equatorial mount we are building is called a barn door type. These are simple mounts. Basically it is two pieces of wood, connected at the ends with a hinge, that allows the wood to open and close like a book. The bottom board is aligned to the star Polaris. The top board moves upward via a drive screw that keeps it in sync with the movement of the stars. A camera mounted on the top board tracks the stars with enough accuracy to make long exposures possible.

The advantages this equatorial mount has over previous designs are: 1) simple electronics that control a stepper motor, and 2) a flexible shaft that simplifies an already simple design.

CONSTRUCTION

Start by cutting two pieces of 3/4 inch thick lumber (plywood or pressboard is fine) 8 inches by 12 inches into "T" shaped pieces of wood, see **Figure 15-2**. Save the scrap wood cut from these pieces we will use them later. Across the 8 inch dimension, join the two pieces of wood with a pair of 3 inch brass hinges. The hinges are attached so that the two wood pieces can open and close like a book, see **Figure 15-3**. On one board, (that will become the bottom) measure exactly 11 3/8 inches from the center pin of the hinges. Drill a hole here, centered side to side, that will accept a 1/4-20 T-nut. Hammer a T-nut into this hole from what will be the inside when the boards are folded together. This hole is where we will place the 2 1/2 inch drive screw.

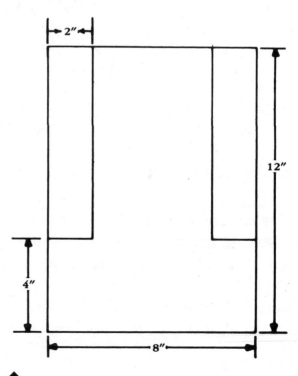

Figure 15-2. A wood plank for a mount.

Drill another hole on the bottom board. Center this hole on the left or right hand side of the 8 inch section of the "T". Hammer another T-nut from the inside into this hole. This hole is where we secure the tracker to a tripod. Note: the 1/4-20 thread of the T-nut is standard for camera tripods.

Drill a hole on the top board about 6 inches from either end, centered side to side. Hammer another T-nut in from the inside. This hole is for securing the camera to the mount.

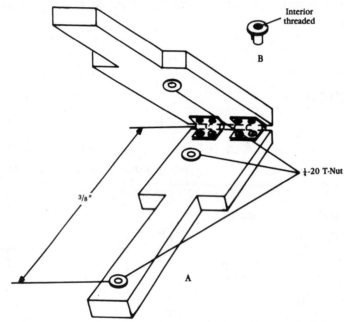

Interior threaded

B

¼-20 T-Nut

3/8"

A

Figure 15-3. a) The placement of hinges, and; b) a perspective view of a t-nut.

CAMERA ARM

The camera arm is made from one of the scraps of wood. It measures about 2 inches wide by three inches long, see **Figure 15-4**. Drill a hole on one end through the width of the wood. On the other end drill another hole through the thickness. Secure an "L" shaped bracket to the T-nut on the top board of the mount. Pass a 2 1/2 inch bolt through the other side of the L

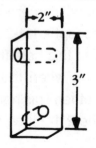

2"

3"

Figure 15-4. The camera arm.

bracket through the hole on the width of the camera arm. Secure it with a wing nut. Pass a 2 inch 1/4-20 bolt through the other camera arm hole. Secure it with a plain nut. Screw a wing nut onto the protruding bolt as shown. The camera body is now placed onto the same bolt and locked into place by tightening the wing nut up to the base of the camera. Adjusting the wing nuts at each end of the camera arm allows you to position the camera at any desired angle.

THE CIRCUIT

Although the circuit is simple it is quite functional, see **Figure 15-5**. There is nothing critical about the circuit; it can be hardwired on perf board, although the PC board helps (see **Figure 15-6** and **Figure 15-7**). The circuit uses three switches, power on-off, fast-slow, forward-reverse. The switches can be located on the PC board as I have done, or placed on a remote switch panel. The stepper motor is

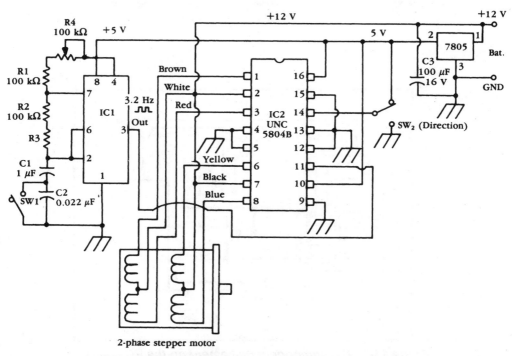

Figure 15-5. A schematic of the stepper motor drive.

connected to the circuit using PC board wire terminals. These screw type terminals make it easy to connect the stepper motor to the circuit. In addition, if you have a different stepper motor than the one specified, you can easily try it out in the circuit to see if it works.

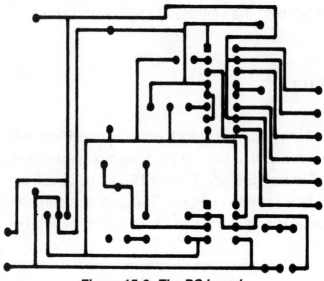

Figure 15-6. The PC board.

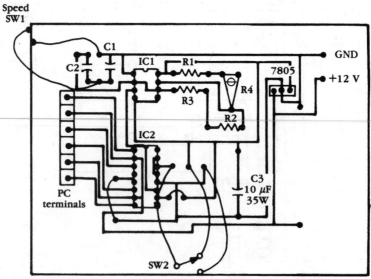

Figure 15-7. The component placement on the PC board.

When the circuit is finished, we must check and adjust its speed before attaching it and the stepper motor to the equatorial mount. With the speed switch set to slow mode, time one revolution of the stepper motor. What we want to achieve is one revolution per minute. Adjust potentiometer R4 to speed up or slow down the stepper motor. The closer you get to one revolution per minute the more accurate the equatorial mount will be. You may want to attach a wire indicator arm to the stepper motor shaft with clay or putty to make it easier to time a revolution. When you have the speed adjusted to one revolution per minute, check out the forward-reverse switch and also the fast speed setting. The use of these controls will be explained later on.

It's time to secure the circuit board and stepper motor to the mount. The circuit is attached to the top side of the top board (see **Figure 15-8** and **Figure 15-9**). I used a few machine screws and aluminum spacers.

Figure 15-8. The side view of the equatorial mount.

The stepper motor is secured to the bottom side of the bottom board, see **Figure 15-10**. It is positioned on the opposite side from where the T-nut was placed for the tripod. I lowered the stepper motor from the bottom board to decrease the stress placed on the flexible shaft. Using three 2 inch x 1 1/2 inch x 3/4 inch wood

Figure 15-9. The top view of the equatorial mount.

spacers, made from the scrap wood, secure each piece of wood to the wood above it with hot glue. If you don't have a hot glue gun, epoxy will work just as well. Drill two small holes on the bottom board on each side of the wood blocks. Secure the stepper motor by threading wire through the holes and around the stepper motor. Twist and tie off. The assembly is then made more secure by hot gluing (or epoxy) the wire to the wood blocks and stepper motor.

FLEXIBLE SHAFT

All that remains is to attach the flexible shaft. The couplers attach the stepper motor and 2 1/2 inch long driver bolt to the flexible shaft. The coupler's inside diameter is 1/4 inch, which is fine for both the motor shaft and bolt. The flexible shaft itself is about 1/8 inch diameter, so you need to use a small piece of plastic tubing over the shaft to increase its diameter before inserting it and locking it into the couplers. In a pinch, wrap electrical tape around the flexible shaft to build up its diameter.

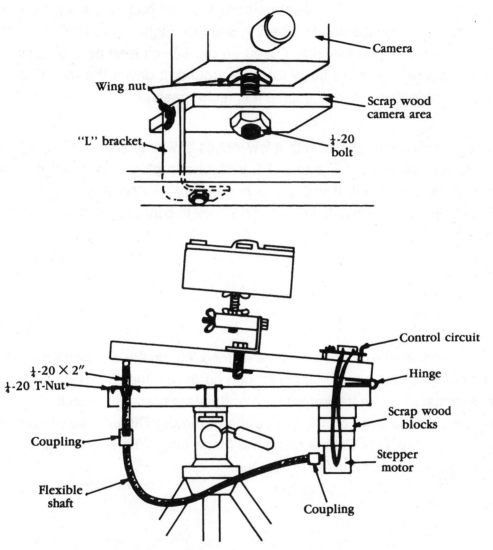

NOTE: Tripod is actually behind stepper motor:
shown this way for illustration

Figure 15-10. The placement of the equatorial mount on the tripod and flexible shaft.

BATTERY

The circuit and stepper motor operate from a 12 volt battery supply. You can use any 12 volt battery capable of delivering 500 mA. Eight 1.5 volt "D" cell batteries strung in series will do the trick. I used an old 12 volt lead acid car battery. Although this is a little more bulky to carry around, it drives the mount for hours without running down, and of course it is rechargeable.

Wire a 1/8 inch phono socket with a few inches of wire to the power supply terminals. Solder five feet of wire to a 1/8 inch phono plug to connect the battery power-pak to the circuit. If you use a car battery as I have, you can purchase spring-loaded claw terminals from Radio Shack to connect the plug to the battery terminals.

USE

To use the equatorial mount it must be aligned. Attach the equatorial mount on to the tripod. Point the equatorial mount due north with the hinges on the left side. Now we must align the tracker with the north star Polaris. You can find Polaris by using the pointing stars on the front edge of the Big Dipper, see **Figure 15-11**. Use the tripod to sight Polaris along the hinge pins. With Polaris sighted in this way fasten the tripod head firmly in place.

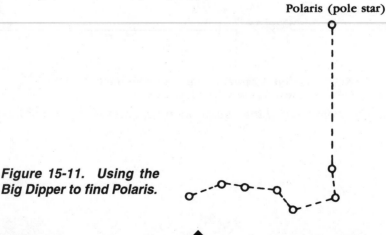

Polaris (pole star)

Figure 15-11. Using the Big Dipper to find Polaris.

You may put a small diameter tube along the hinge pins to use as a sighting tube for Polaris. I used a 2 inch piece of a plastic straw.

At this point the equatorial mount is aligned and you're ready to go. **Figure 15-12** shows the completed setup. The driver screw should be all the way down (don't bind the drive screw into the wood, leave 1/8" clearance). The controls on the circuit should be set to slow, with the direction counterclockwise to slowly rise the upper board when the stepper is started. Align the camera to any area or star you want to photograph. Now start the stepper motor. The stepper motor will turn the drive screw at one revolution per minute. This slowly lifts the upper board (and camera) in sync with the movement of stars.

I would advise testing the equatorial mount with 1, 2, 3, 4, 5 and 10 minute exposures. Work towards longer exposures as you gain experience. You may also want to use a telephoto lens on the camera. The computer simulation on the tracker shows it to be accurate enough to use a 200 mm lens for 30 minutes.

Figure 15-12. The equatorial mount on the tripod.

When the exposure is complete stop the stepper motor, reverse direction and put it in fast mode to quickly lower the board to reset the equatorial mount for another photo.

ANOTHER WAY TO USE THE MOUNT

Some equatorial mounts will track more smoothly with the board moving downwards instead of up. No problem. To use the equatorial mount like this, have the hinge pins to the right when you align the tracker with Polaris. The mount should be started with the boards opened. Set the controls to turn the drive screw clockwise. Now when you start the stepper motor the boards will slowly close.

FILM SPEED & APERTURE

You have a large variety of films available. I would advise using a fast ASA color print or slide film, like a 1600 or 3200 ASA speed. The aperture setting on the camera controls the amount of light entering the camera. The smaller the f-stop the more light that enters. You want to set the aperture to its widest opening (lowest stop number) usually around 2 or 2.4.

With the camera set at its largest aperture setting, the lens may not produce the sharpest picture. If you get unsharp pictures, you may need to stop down to get the sharpest picture possible.

It's also a good idea to start each roll of film with a few standard photographs, this way the photolab that does the developing will have frame markings to use. Otherwise, they may end up inadvertently cutting some pictures in half. Also tell the photolab to print all pictures. They may take your star photographs as errors and not print them.

◆

LIGHT POLLUTION

If you live in or near a large city you will encounter a common problem, light pollution. I live in NYC and the photos I took of the sky looked like dirty brown soup with a couple of stars.

You have two options: take the tracker away from the city lights to shoot or purchase a light pollution filter. Light pollution filters are available from Orion and Lumicon, see parts list.

These filters block out the parts of the visual spectrum that are emitted by street lamps and other artificial light sources, at the same time allowing transmission of the rest of the visual spectrum for your pictures.

HOMEMADE LIGHT POLLUTION FILTERS

If you don't want to make the initial investment in a commercial light pollution filter you can try your hand at making one. You are restricted to using black and white film with this simple filter. Use an extreme red filter, like a #25 or #29 filter.

PARTS LIST

	ITEM		SOURCE
IC1	555 Timer	#276-1723	Radio Shack
IC3	7805 Voltage Regulator	#276-1770	Radio Shack
R1	10K Resistor	#271-1335	Radio Shack

	ITEM		SOURCE
R2	100K	#271-1347	Radio Shack
R3	47K	#271-1342	Radio Shack
R4	100K PC Mounter Pot.	#271-285	Radio Shack
C1	1 μF	#272-996	Radio Shack
C2	.022 μF	#272-1066	Radio Shack
C3	100 μF	#272-1016	Radio Shack
SW1		#275-634	Radio Shack
SW2		#275-635	Radio Shack
IC2	UNC-5804B Stepper Controller		Images Company
IMMT	Stepper Motor		Images Company
PCSTP	PC Board		Images Company

CHAPTER

HOLOGRAPHY

This topic is a two part series contained in two chapters. In this chapter will shall take a look at some basic concepts of holography and build a laser system capable of producing holograms. Holography can be a very simple process, which you will discover for yourself in part two. You don't have to be an expert in the technical areas to shoot and develop first rate holograms.

Holography, like photography, is a technique that produces an image on film. Holography, however, records all the visual information of a three-dimensional image, including depth. Subsequently, this allows you to view the original scene from many different angles. In essence, you can look around objects in the hologram.

The roots of holography trace back to 1947 when Dr. Dennis Gabor developed holography in the hopes of increasing the resolution of electron microscopes.

The evolution of the technology has been slow. There are many holographic frontiers to conquer and technologies to develop. This is not to say that holography hasn't advanced to a point where you can't use the existing technology for purely artistic pursuit; it has.

HOLOGRAPHY VERSUS PHOTOGRAPHY

We might best understand the process by which holography records three-dimensional images by comparing it to photography.

Photography creates a two-dimensional image on film. The film image is called a negative. The photographic image is a single unchangeable viewpoint. The third dimension (depth) is collapsed onto the plane of the film. **Figure 16-1** is a drawing of a simple box camera. The image formed by the lens onto the film is a real image. Lighting for photography can come from any number of common light sources; sun, electric lights or flash tubes.

Looking at the subject in a photograph from an angle just creates a foreshortening of the flat image, the image remains at the one viewpoint alone.

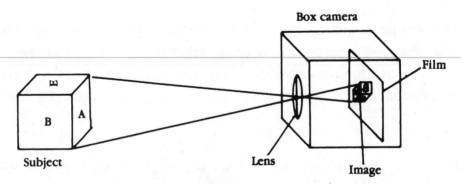

Figure 16-1. Image formation in a simple camera.

Holography does not record the subject's image the same way a camera does. Holography records the interference pattern of light generated from a reference beam and reflected light from the subject (object beam). The light source required must be monochromatic (single light frequency) and coherent (wavelengths in phase). A helium-neon (HeNe) laser fits the bill.

Figure 16-2 illustrates a typical split-beam holographic setup. It is possible to produce holograms using a single beam. In fact, the holograms presented in part two of this article are single beam setups, but in an effort to present a diagram that clearly illustrates the interference pattern created, a split beam setup is better.

There is no negative to produce pictures as in photography, the original film exposed and developed *is* the hologram.

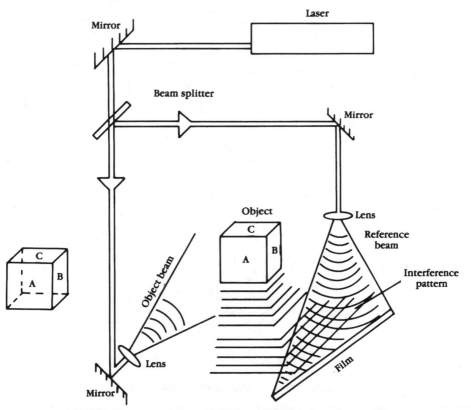

Figure 16-2. Interface pattern formation in holography.

When the finished hologram is viewed, it is a true, three-dimensional image of the subject. **Figure 16-3** demonstrates the parallax of the hologram. The vertical parallax is from top to bottom, and the horizontal parallax is from left to right. Parallax is a term referring to the viewable angles of the subject in the hologram.

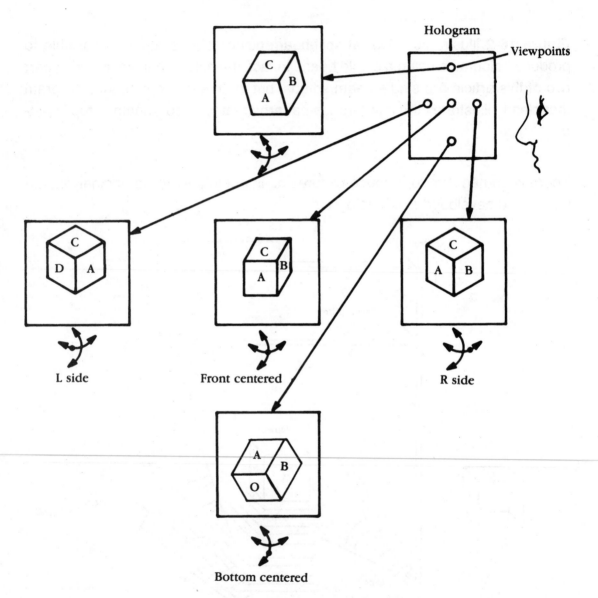

Figure 16-3. Parallax in a hologram.

REDUNDANCY

If a hologram is broken into small pieces, the entire image would still be viewable through any of the broken pieces. This is easier to comprehend if we look at **Figure 16-3** again. Imagine that the holographic film becomes a window with memory when it is exposed. So any object behind the window, from any viewpoint, is faithfully recorded, so much so that if you covered the hologram with a black piece of paper with a peep hole in it, you could still view the entire subject through the peep hole as if you were looking through a window. Where you placed the peephole on the window (hologram) determines from which perspective you would see the subject.

TWO TYPES

There are two basic types of holograms; reflection and transmission. The reflection holograms can be viewed with normal white light (also called white light reflection). This is the type of hologram we will produce. Transmission holograms require a monochromatic (single frequency) light source for viewing. Typically a laser is used for illumination.

LASERS

The first step in producing holograms is acquiring a suitable light source. We already established that a helium-neon (HeNe) laser fills the bill, so you need to decide either to build or buy one. Whatever your choice, the laser must meet certain specifications to be capable of producing holograms. The most important specification is that it operates in TEMoo Mode. The power output is our next consideration.

The power of the laser is directly related to exposure time. Shooting a hologram with a 1 milliwatt laser will require a longer exposure time than a 2 milliwatt laser. For beginners, either a 1 or 2 milliwatt laser is suitable, and will keep your startup costs down. Later if the holography bug bites you, you can upgrade to a more powerful laser.

A 1 milliwatt laser tube costs about $30.00; 2 milliwatt about $55.00. Laser tubes can be purchased from Images Company, see parts list.

LASER LIGHT

Before we get into the nuts and bolts of building our laser, let's first understand what a laser is. The following is a simplified explanation of laser function.

The word laser is an acronym that stands for Light Amplification by Stimulated Emission of Radiation. Albert Einstein first theorized on stimulated emission of radiation. When an atom absorbs energy, an electron in one of its shells jumps up to a higher energy level. An atom in this state is said to be excited. When the electron spontaneously jumps back down to its lower energy level it will emit a photon of radiation. This photon is equivalent in energy to the difference of the two energy levels that the electron jumped. If this emitted photon happens to collide with another excited atom, it will stimulate that atom to release a photon. The stimulated photon will have the same frequency and be in phase with the original photon. This is the Stimulated Emission of Radiation part of the laser. To achieve Light Amplification, we need to produce a "population inversion". What this means is that we have a large number of atoms in the excited state. As a few of the atoms begin to emit photons spontaneously, they create an avalanche of photons through stimulated emission. The laser has mirrors at both ends that form an optical cavity. The photons emitted bounce back and forth between the mirrors, producing a beam of light (through stimulated emission) that is monochromatic (single frequency) and in phase.

One of the mirrors that make up the optical cavity is less than 100% reflective. It is this mirror that allows a small percentage of light to pass through, which is the laser beam.

LASER POWER SUPPLY

A commercial power supply for either of these tubes costs about $80.00. We can build one for less than $40.00.

CAUTION: The laser power supply is a high voltage device and as with all high voltage devices must be handled cautiously. The power supply should never be operated without a load connected to its output. Failure to do so will cause arcing and electrical discharge that may damage the power supply.

Figure 16-4 is the schematic of the power supply. The circuit is kept simple by the use of the high voltage transformer T2. The two items critical in the circuit are T1 and Q2.

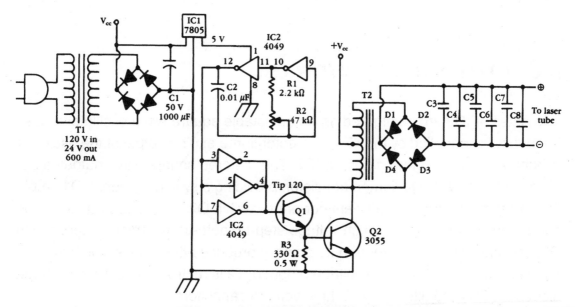

Figure 16-4. A schematic of the laser power supply.

The transistor Q2 must be adequately heat sinked for long continuous operation. I typically use plastic enclosures for high voltage circuits but I made an exception with this unit. In order to provide greater heat sinking capacity for Q2, I attached Q2 to the wall of the metal enclosure. To increase the heat sinking capacity further I added some metal strips for additional mass.

The need for adequate heat sinking is to allow the circuit to power the laser continuously for hours without any degradation. This is critical when you want to shoot holograms.

The transformer T1 provides the power to the circuit. T1 is specified in this circuit at 120 VAC/24 VAC, .6 amps. This transformer is ideal for powering all laser tubes from .25 milliwatts up to and including 1.5 milliwatts. Larger transformers (higher output current) can be substituted for T1 in the circuit to power larger laser tubes but you will have to provide greater heat sinking to transistor Q2.

The high voltage capacitors and diodes are assembled on a separate piece of PC board. This isolates the high voltage from possibly arcing over to the oscillating section of the circuit.

HOW THE SUPPLY WORKS

Two gates of IC2 are configured as an adjustable square wave oscillator whose frequency is controlled by the 47 Kohm potentiometer. The output of the oscillator is buffered by three other gates on IC2. The buffered square wave output is applied to Q1, an NPN Darlington transistor, providing a switching signal. Q1 amplifies the signal to provide sufficient current to switch Q2, a TIP3055 transistor, on and off. Transformer T2 is a high voltage step-up auto-transformer, in series with Q2, and the switching current on its primary produces the required high voltage on the secondaries. The high voltage output of T2 is rectified by four high voltage diodes D1-D4, and filtered by six high voltage capacitors.

◆

CONNECTING THE POWER SUPPLY TO THE LASER

All helium-neon laser tubes require a ballast resistor to limit the current flowing through the tube. When you purchase the laser tube be sure you also purchase the ballast resistor. Most ballast resistors range between 50 kilohms and 200 kilohms with a 3-5 watt capacity. The resistor generally connects to the anode (+) side of the laser tube.

It is sometimes hard to recognize the anode (+) and cathode (-) terminals on the tube. Sometimes the positive terminal is marked with an "A", "+" or a small red dot. The negative terminal is sometimes marked with a "C" or "K". The cathode may also be identified by a small metal tube on one end that was used to fill it with gas.

Most tubes today are hard sealed. The metal terminals on the end are also mirror mounts. The mirrors are precisely aligned to form the optical cavity. This makes it a bad idea to solder wires directly to the terminals because the heat of soldering may throw the mirrors out of alignment. Some companies sell beryllium copper spring clips that can clip onto the terminal. In a pinch, you can use 1/4 inch fuse clips from Radio Shack (# 270-739), see **Figure 16-5**. The clips are attached to a

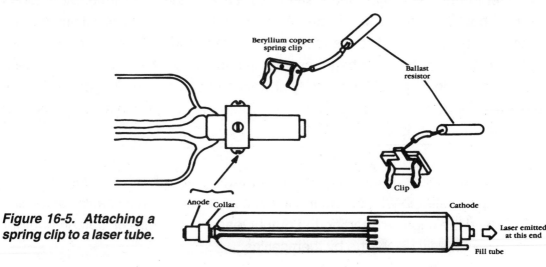

Figure 16-5. Attaching a spring clip to a laser tube.

bakelite base. Leave the clips attached to the bakelite and cut the bakelite in half so you have a clip on each side. Then bend the clip open a little to adjust it to the size of the laser terminal. Solder your lead wires to the clip and then place clips onto the laser terminals.

If you are powering a laser head (a laser tube enclosed in a housing), identify the polarity by the color of the leads coming out of the housing.

TESTING & CALIBRATING THE LASER POWER SUPPLY

NEVER operate the laser power supply without a load connected to its output.

As stated before, all laser tubes require a ballast resistor to limit the current flowing through the tube. Be sure you purchase and have the resistor in series with the laser tube before you turn on the power supply. Failure to do so may cause permanent damage to the laser tube.

To calibrate the power supply to your laser tube use a VOM, see **Figure 16-6**. Set the VOM to read milliamps and place in series with the laser tube. Most tubes require about 5 milliamps to operate. Turn on the laser supply and adjust potentiometer R2 until the VOM reads the proper current and the laser is producing a steady beam. Use the laser beam as your final guide, as the tube may require a little more current to produce a steady unwavering beam. Allow the power supply to operate for 30 minutes or so. During this time the components will break in. If the power output drops, readjust R2.

When you are finished adjusting the power supply, turn it off and complete the power supply assembly. Remember, the capacitors will retain a charge for awhile. Either bleed the caps by shorting the output leads with a wire or wait an hour for them to completely discharge by themselves.

◆

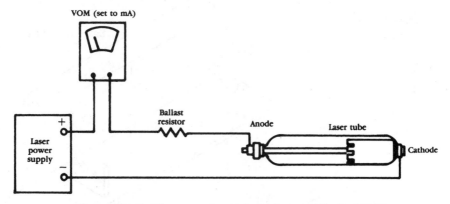

Figure 16-6. The supply connections with the VOM.

LASER TUBE HOUSING

After you have finished checking out and adjusting the power supply, we need a suitable mounting and enclosure for the laser tube. I used a laser head that has the laser tube enclosed.

Some simple mounting ideas are as follows. Locate plastic snap-in clamps (used to mount capacitors) with the same diameter as your laser tube. Mount them to a wood base and secure the laser tube in them.

Another way is to get a piece of lumber 4 x 3 x 3/4 inches thick. Drill a hole in the center of the wood the same diameter as the tube or laser head. Large diameter cutout drills are available in hardware store to drill doorknob openings in doors. If they don't have the exact diameter you need, select the closest diameter that is a little larger. After you drill the wood, cut the wood in half so that you have two semicircles on each piece. Mount these pieces to wood as illustrated in **Figure 16-7** with wood screws. On each side of the wood, place a wood screw as shown in the detail drawing. Place the laser in the wood pieces as shown. Using 22-gauge insulated wire, wrap the wire around the wood screw and lash it across the laser. Secure it to the other side using the wood screw in the same manner. Do this with both mounting pieces of wood.

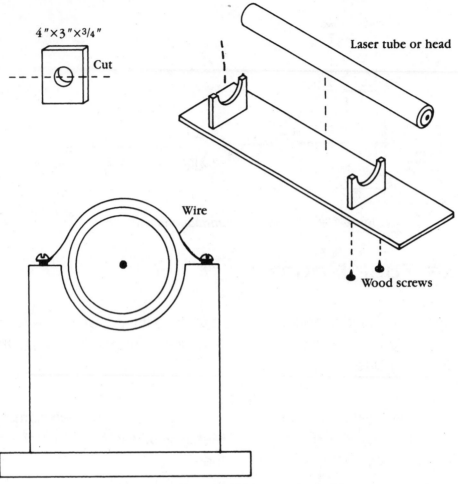

Figure 16-7. The simple laser tube holder.

If you are using an enclosed laser head you're finished. If you are using a standard laser tube, you need to build an enclosure. The reason you need an enclosure is to prevent the glow of light from the tube from exposing the film you plan to shoot. Paint the inside of the enclosure black, and don't forget to drill a hole in one end of the enclosure to allow the laser beam to pass through.

On another laser project I used a rectangular aluminum pipe for an enclosure. I encountered a problem you should be aware of. Because of the close proximity of the tube to the walls of the enclosure, the high voltage arced to the tube walls. I solved this problem by masking the inside of the pipe with electrical tape.

LASER SAFETY

Never look directly into the beam of a laser. An unspread beam from a small .5 milliwatt laser is well above the ANSI standard for eye safety.

PARTS LIST

	ITEM		SOURCE
	Bridge Rectifier	#276-1171	Radio Shack
IC1	7805 Voltage Regulator	#276-1770	Radio Shack
IC2	4049 Hex Buffer	#276-2417	Radio Shack
Q1	TIP 120 NPN	#276-2068	Radio Shack
Q2	3055 NPN (TO-220 case)	#276-2020	Radio Shack
C1	1000 µF 50V	#272-1047	Radio Shack
C2	.01 µF	#272-1065	Radio Shack
R1	2.2K	#271-027	Radio Shack
R2	47K Potentiometer	#271-283	Radio Shack
T1	24V .6 Amp Transformer		Mouser Electronics
T1	24V .6 Amp Transformer		Images Company
T2	10 KV Transformer		Images Company
D1-D4	10 KV 10 mA Diodes		Images Company
C3-C6	6 KV .002 µF Cap		Images Company
	HeNe Laser Tube 1-2 mW		Allegro Elect. Sys.

CHAPTER

---◆---

HOLOGRAPHY II

Our last project was a HeNe laser. Now we will continue with the construction of the components needed to shoot holograms.

WHERE TO SHOOT

The first step is deciding where you can setup the equipment to shoot. The area must be quiet (no vibrations) and dark (no light). There are safelightes available so you do not have to work in complete darkness and we will cover them later. I usually set up my holographic equipment on the floor to reduce vibration. A concrete floor in your basement is perfect, but a bathroom or bedroom floor may be OK, too.

ISOLATION TABLE

Holograms are very sensitive to vibrations. Vibrations so subtle that you can not feel them can prevent the hologram from forming. Because of this, holographers use an isolation table. The isolation table is designed to dampen as much vibration as possible. The table we will use is simple and portable. The table can be setup or stored in less than a minute.

The table consists of three components; a small piece of carpet, a small 12-18 inch diameter inner tube, and a metal plate, see **Figure 17-1** and **Figure 17-2**. The carpet should be large enough so that the inner tube can lay on it without hanging over the edge. The inner tube has just enough air for it to be filled but still remain very soft. In other words you could squeeze the sides of it together easily. The top metal plate is our working area and should be about the same size as the carpet. The metal should be thick enough so that it doesn't flex when components are place on it. The plate I use is 3/16 inches thick. If you cannot get a metal plate you can use 3/4 inch or thicker plywood, with sheet metal adhered to one side.

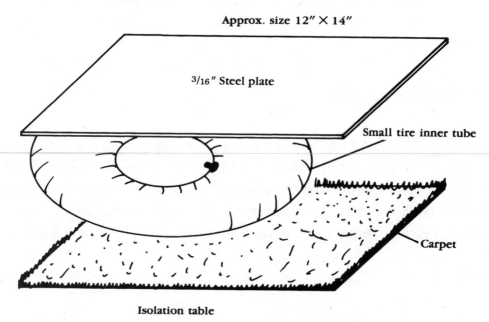

Approx. size 12″ × 14″

3/16″ Steel plate

Small tire inner tube

Carpet

Isolation table

Figure 17-1. The simple isolation table.

Figure 17-2. The isolation table set up to shoot holograms.

OPTICAL COMPONENTS & MOUNTS

We will keep our optical components to the bare bones minimum; one mirror lens. The mounts for the optical components are steel plates that measure 1 x 4 1/4 x 1/16 inches thick. The plates can be assembled in a variety of useful configurations using small bar magnets. I advise buying a minimum of 4 plates and magnets. The most basic configuration looks like an upside-down "T". Putting the plates together is simple, see **Figure 17-3**. The unit is more stable than it might appear.

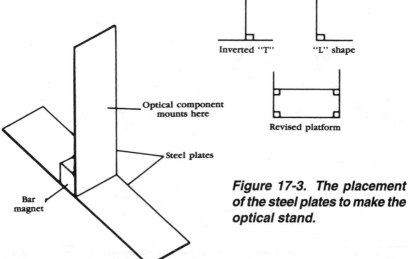

Inverted "T"   "L" shape

Optical component
mounts here

Revised platform

Steel plates

Bar
magnet

Figure 17-3. The placement of the steel plates to make the optical stand.

When using a metal table you can eliminate the bottom portion of the "T" and use a magnet directly on the table top.

SECURING OPTICAL COMPONENTS

The mirror lens is a spherical front surface mirror attached to a bar magnet using epoxy glue, see **Figure 17-4**. Be careful not to get any glue on the front surface.

With the component so mounted, it easily attaches to the side of the inverted "T", and is adjustable through a full range of motion, see **Figure 17-4**. This makes aligning and directing the laser light easy. The spherical mirror has a short focal point, see **Figure 17-5**. When the laser passes the focal point it spreads rapidly, allowing you to illuminate the entire photographic plate with laser light.

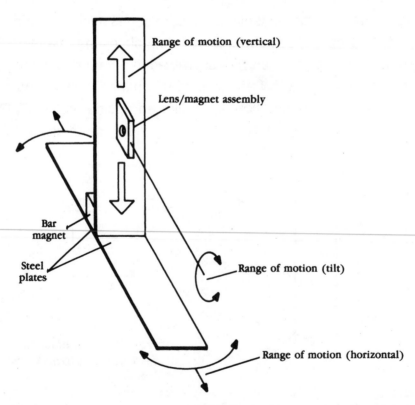

Figure 17-4. The placement of the optical component on the stand.

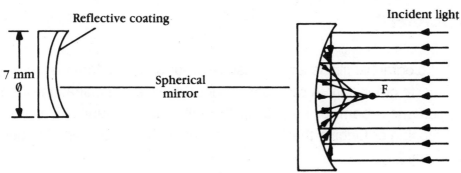

Figure 17-5. The spherical mirror.

FILM

There are many kinds of film you could use to shoot holograms. My recommendation is to start with 2.5 inch square 8E75 glass plates. Using glass film plates simplifies the setup, procedure and of course the film holder itself.

FILM PLATE HOLDER

The film holder consists of two office binder clips with a magnet glued to each one, see **Figure 17-6**. The binder clips are available from any store that sells office supplies.

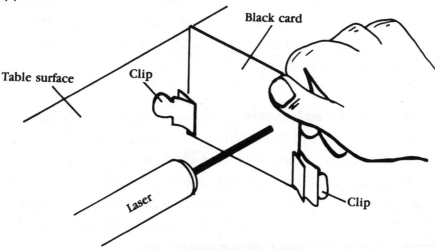

Figure 17-6. The shutter card.

SAFELIGHT

The 8E75 holographic film is least sensitive around 500 nanometers (nm) wavelength, equal to green light. A safelight provides sufficient illumination when working in a darkened room for setting up the table and/or during film processing without exposing or fogging the film. I recommend buying a safelight specifically for holography.

SHOOTING YOUR FIRST HOLOGRAM

First step, turn on your laser. The laser must be given ample time to warm up and stabilize before you shoot. The laser must be operating 20-30 minutes before you expose a plate.

CHOOSING AN OBJECT

The first object you select to holograph should be smaller than the plate and preferably a light color; white, off white, silver or metallic finish. You can shoot larger and darker items later after you gain some experience, but to start with, holograph something that will show up brightly. The object should also be rigid, something that won't flex or bend or move during the exposure. For my first object I chose a small white sea shell.

The object must be secured to prevent it from moving or rocking during the exposure. One of the easiest ways to accomplish this is by putting a small ball of clay on the table where you're placing the object. Push the object into the clay. Instead of clay, you could also use "Fun-Tak". Fun-Tak is a blue colored adhesive material with the consistency of clay. It can be pulled apart and pushed together again, rolled into balls, whatever. The material is reusable and because of its strong adhesive properties is worthwhile to purchase.

◆

Another way to secure objects is to glue the object to a magnet and position them with a steel plate behind the film plate.

We will shoot a hologram that is viewable in white light. In addition to the materials already mentioned you will need a white cardboard card the same size as a film plate, and a black card with two binding clips to block and unblock (shutter) the laser beam, see **Figure 17-6**. Set up the holography table as illustrated in **Figure 17-7**. Position the white card, using the film holder clips, where the film plate will be located. Adjust the mirror so that the beam reflected from the mirror is spread evenly on the white card. Remove the white card, leaving the magnet binding clip(s) in position. The laser is now illuminating the object you are shooting. Position the object close to where the film plate will be. Look at the object from the laser side; this is what your finished hologram will look like. Make any adjustments you want to the object to holograph it in the best position possible. Block the laser beam with the black shutter card.

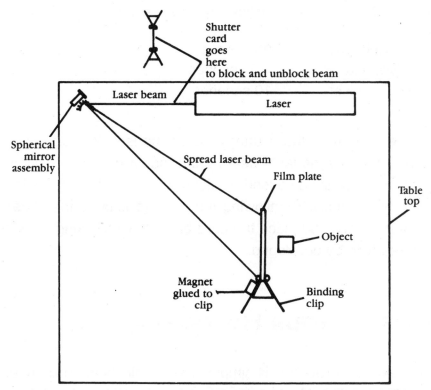

Figure 17-7. The top view of components on the isolation table.

With all the lights off, remove a film plate from its light tight box. Try to hold the plate by its edges. Close the box and turn on your safelight. The film plate appears transparent, but there is an emulsion on one side. We want to place the plate emulsion side toward the object as this will give a better hologram. Touch the plate with a moist finger on each side at the corner. The emulsion side is sticky. Don't worry if you can't identify the emulsion side when you first start out. You will produce a hologram with the emulsion faced either way.

Put the plate in the binding clip(s) that were holding the white card. Leave the setup alone for a minute. This will allow any vibrations to die down.

MAKING THE EXPOSURE

The exact exposure time varies with the intensity of the laser light and the sensitivity of the film. You can use the following times as a start.

1 milliwatt laser	20 seconds
1.5 milliwatt laser	15 seconds
2 milliwatt laser	10 seconds

To make the exposure lift the shutter card off the isolation table, but keep it a position that still blocks the laser beam. Hold the card in this position for 30-60 seconds to let any vibration caused by lifting the card off the table to die down. Then lift the card completely, allowing the laser to expose the plate. After the exposure time has elapsed place the card back down blocking the laser beam. The plate is ready to be developed.

DEVELOPING YOUR HOLOGRAM

Although developing holograms is simple, the chemicals are poisonous and can be absorbed through the skin. Since I use my hands to move the plate from tray to tray I always wear rubber gloves when developing holograms; you should too. You need three plastic trays large enough to hold a film plate. The trays are arranged as in **Figure 17-8**. The first tray holds the developer, second tray water and the third tray contains the bleach. The chemistry involved is simple, but it is easier to buy a kit than to procure the chemicals separately on your own. The kit includes pre-measured packets of chemicals that you just dump into a quart of water to make up your stock solutions. Images Company sells an excellent developing kit (JD-2) for $12.00. Mix the chemicals according to the directions in the kit.

(Note: the instructions for the kit require that the chemicals be mixed in 1 litre of water, for our purposes you can substitute a measured 1 quart of water for 1 litre without any harmful effects.)

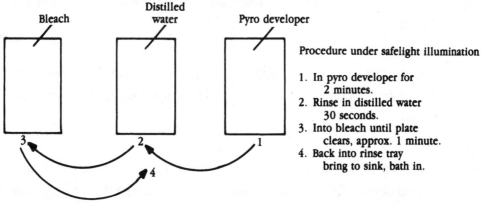

Figure 17-8. The procedure for developing holograms.

Use a safelight for illumination during development. The developer is made from equal parts of two stock solutions labeled A and B. Mix equal portions of the solution in the tray, just enough to cover the plate when you place it in the tray. The developing solution has a short lifetime, so mix it just before you're ready to develop holograms. The plate is placed in the developer, emulsion side up, for two minutes. Gently rock the tray back and forth to keep fresh solution in contact with the plate. The plate will gather density and may appear to turn completely black. Don't worry; that's normal.

Remove the plate from the developer and place it in the water for 30 seconds. This step isn't mandatory, but it will extend the life of the bleach so you can reuse it.

After the time has elapsed in the water, place the plate in the bleach. Rock the tray gently back and forth as before. Keep the plate in the bleach until it becomes completely clear again, usually about one minute. After the plate clears, the emulsion is light safe, so you can turn on the room light.

Remove the plate from the bleach and put it back into the water tray. Bring the tray to a sink, run tap water at about room temperature and place the tray under the running water for five minutes. Afterward, remove the plate, stand it vertically against a wall and allow it to dry. The holographic image will not be visible until the hologram is completely dry.

Plates sometimes dry with water spots on them. You can dip the plate in Kodak Photo-flo solution (mix according to direction) after the final rinse to prevent water spots from forming.

Some holographers wipe the plate with a squeegee to remove excess water and thereby speed up drying. Others use a hair dryer to shorten the drying time. If you use a hair dryer, set it on warm or low, or you may damage the hologram from excessive heat.

VIEWING YOUR HOLOGRAM

After you have exposed and processed your holographic plate, it's time to take a look at your hologram. First, let the plate dry because if it's still wet you probably won't see anything.

The hologram we made is a white light reflection, and as the name implies, it is viewable in white light. The best type of illumination for this hologram is a point light source. The sun is a perfect example of such a source.

Tungsten halogen lamps are an excellent light source. Incandescent lamps can be used, but the image quality isn't as good. When using an incandescent lamp, notice that when increasing the distance of the hologram from the bulb, the image appears sharper. This happens because the lamp becomes more like a point light source as the distance increases.

To improve the quality of the playback image, put a black sheet of paper behind the hologram. To make this effect permanent, spray paint the back of the hologram black. Do this only with reflection holograms you want to display because once it's painted, you can't use it to make copy holograms.

REAL AND VIRTUAL IMAGES

There are two types of images we can view with our white light reflection hologram, real and virtual. The most common example of a virtual image is the image that is reflected in a mirror, see **Figure 17-9**. To the observer, the reflected rays that appear to come from the virtual image do not actually pass through the image. For this reason, the image is said to be virtual.

The parallax and perspective of the virtual image in a hologram is observed to be correct. As you move your head from side to side

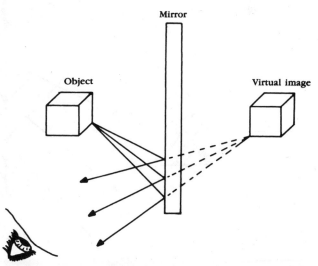

Figure 17-9. The virtual image reflection in mirror.

or up and down, the three-dimensional scene changes in proper perspective as if you were observing the real physical scene, see **Figure 17-10**. The virtual image is said to be an orthoscopic image, meaning true image.

In contrast to the virtual image is the real image. The real image in our reflection hologram can be observed simply by flipping the hologram around. The real image has some peculiar properties. The perspective is reversed. Parts of the image that should appear in the rear are instead in the front and vice versa. If you move your head to the right, the image appears to rotate in proportion to your movement and you see more of the left side, not the right side. The brain perceives this paradoxical visual information causing the image to swing around. The real image is said to be pseudoscopic, meaning a false image. The real image is useful when you want to make copies of holograms.

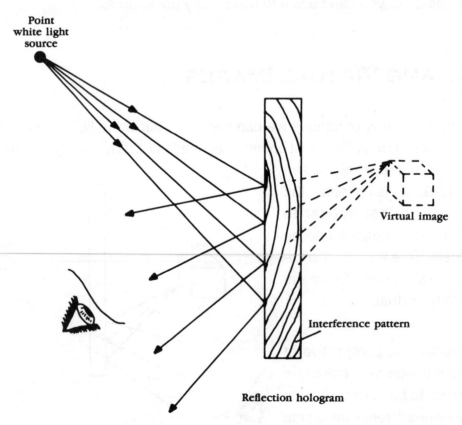

Figure 17-10. Virtual image reconstruction in reflection hologram.

TROUBLESHOOTING

Not all the holograms you shoot will be perfect. The following list will help locate the problem when you encounter some fault with the hologram's image.

No Image. A hologram without any image can be a very frustrating problem. Even though it might seem as if you don't have a starting point to begin evaluating the problem, you do. During development, did the hologram go black or did it gain some density. In the latter case, the most probable cause is that the film plate or model moved during exposure.

Parts of Hologram Missing. This relates to the first problem. If you are shooting a hologram that contains more than one object, each object must be secure. If an object moves even slightly during exposure, it will not develop a holographic image.

Dark Spots. This is also caused by slight movement.

Image Has Bands of Light and Dark. The object moved slightly during exposure.

Faint Image. A faint image can be caused by either overexposing or underexposing the film. If the plate went black almost immediately after being placed in the developer it is overexposed. If the plate is still light after two minutes of development it is underexposed. Another possible cause is the tap water used in making your stock solutions contained chlorides. If possible use distilled or deionized water for your stock solutions.

Image Is Weak and Fades. This is a problem I encountered when I used green plastic filters for my homemade safelight. The safelight was fogging my film during setup and development.

◆

Hɪɴᴛs & Tɪᴘs

1) In the illustrations the laser has been positioned on the isolation table. This is not necessary. You could move the laser off the table without any ill effects.

2) Never place the power supply on the isolation table. The 60 Hz hum from the step-down transformer may cause the table to vibrate making it impossible to produce a hologram.

3) The mixed developer (parts A & B) has a short lifetime and is not reusable. After you successfully produce a few holograms you may want to try batch processing, where you expose a few plates, keeping them in a lite-safe box and developing them all at once or one after another in the same developer solution. I don't advise doing this before you have successfully produced a hologram because if there is something wrong with the set-up, you'll be multiplying your errors with subsequent exposures.

4) Remember the chemicals used to develop holograms are poisonous. Always wear rubber gloves when handling or developing holograms.

Oɴᴇ Mᴏʀᴇ Fᴏʀ ᴛʜᴇ Rᴏᴀᴅ

There are many types of holographic setups that you can accomplish with the equipment you have. Transmission holograms, copy holograms, transmission to reflection, pseudo-color reflection holograms and stereograms are all possible setups on the table.

However, let's try one more setup that shows how powerful holography can be. This is called a dual channel or multiplex hologram. What we are doing is holographing two different objects on one plate. In the finished hologram the image changes from one to the other as your viewing angle changes. As an example, exposure one could be an unopened jewelry box, and the second exposure would show the box opened, revealing an engagement ring.

One method of recording multiple images is to change the angle of the light exposing the plate. We accomplish this by moving the holographic plate as shown in **Figure 17-11**. The rest of the setup is the same as we used in the reflection hologram. The exposure time for each exposure should be about half the time used to record a single hologram.

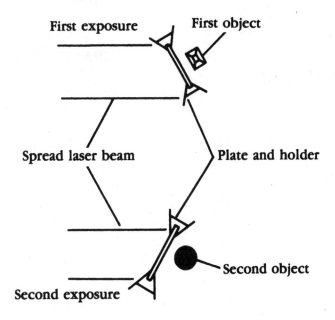

Double channel hologram
(multiplex)

Figure 17-11. The procedure to make a double channel hologram.

PARTS LIST

ITEM		SOURCE
Steel Plate 1 x 4 x 1/16 Inches Thick	#IM-ST	Images Company
Spherical Mirror	#IM-SP	Images Company
Bar Magnet	#IM-M2	Images Company
Large Bar Magnet	#IM-M3	Images Company
30 pcs Agfa 8E75-HD-NAH		Images Company
2.5 Inch Glass Holographic Plates		Images Company
Green Safelite	#IM-LT	Images Company
JD-2 Holographic Developing Kit		Images Company

CHAPTER

KIRLIAN PHOTOGRAPHY

Kirlian photography produces strikingly beautiful photographs of the most common objects. It does so without the use of a camera or lens. The photographs are direct contact prints on film or paper using a high voltage power supply. Before we actually get into making Kirlian photographs let us first look at a little history.

SHORT HISTORY

Kirlian photography is named after Russian researchers Semyon Kirlian and his wife Valentina. Their work with high voltage photography was made known in this country by a book published in 1970 titled "Psychic Discoveries Behind the Iron Curtain" by Sheila Ostrander and Lynn Schroeder. The Kirlians claimed that this type of photography could be used as a medical diagnostic tool, stating that disease in subjects showed up in the photographs as a modified pattern of discharge before obvious symptoms became manifested in the subject. Naturally this generated much interest in this country. More interesting than this is a second claim known as the "phantom leaf" where a small section of a leaf is removed before photographing, but in the subsequent Kirlian photo, the missing section of the leaf appears in the photo.

I have as yet to see these two claims manifested in my own Kirlian photography. I'm not saying they don't exist, just that I haven't seen them.

Although the field of electrophotography is called Kirlian photography in honor of the Kirlians, the Kirlians are not the first ones to experiment with patterns created by electric discharge. The beginning of electrophotography appears to have started much sooner. Georg Christoph Lichtenberg, in the late 1700s, observed pictures made in dust created by static electricity and electric sparks.

USEFULNESS

It appears that there are two main avenues for development. One is to use Kirlian photography for special effects photographs. You can add the equipment and technique to your repertoire of photography. The other is to investigate the potential of this method of photography for its scientific value. You may experiment with Kirlian photography and attempt to verify some of the claims.

DEBUNKING THE MYTHS

Kirlian photography has the potential to become a scientific instrument. Note the operative word "potential" in the last sentence. Despite any claims to the contrary, this type of photography can not, as of yet, be used for any medical diagnostic purposes. Any and all research I have read to date draws inconclusive results. It appears that patterns and "aura" of these electrophotographs are determined by conductivity of the object, pressure on the plate and moisture content of the air. Anyone claiming otherwise is at this point in time perpetrating a fraud.

But herein lies the key for doing legitimate research. By controlling the variable factors mentioned; conductivity, pressure and humidity as well as frequency of discharge, duration of exposure and using a measured and controlled high voltage source, one may begin to lay a foundation of research.

Kirlian photography has already been shown to be useful for some types of non-destructive testing.

COLD ELECTRON EMISSION

The high voltage potential causes a cold electron emission that is rich in ultraviolet light. As far as I know no one has done any UV spectrographic testing of this light. This may be another area for research.

CIRCUIT OPERATION

TR1 is a step-down transformer that brings the line voltage down to 25.2 VAC (see **Figure 18-1**). The power from the transformer is rectified by RECT and C1. IC1, a 7805 voltage rectifier, provides +5V to IC2, a 4049 hex inverter. Two gates off the inverter are set up as an adjustable oscillator using C2, R1 and R2. The oscillator is adjustable, using R2, from approximately 500 Hz to 10000 Hz. The output of the oscillator is fanned into three gates on the 4049 in parallel. The output from the three gates provide sufficient current to trigger Q1, a TIP120 NPN Darlington transistor. Transistor Q1 in turn triggers Q2, a high powered MOSFET.

D1, a 200 volt zener diode, protects the MOSFET from reactive voltage surges by providing a clean path to ground. Resistor R4 provides current limiting through the MOSFET.

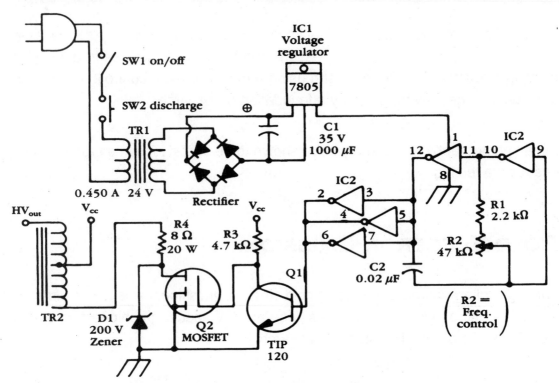

Figure 18-1. A schematic of a Kirlian device.

CIRCUIT CONSTRUCTION

Basically, Kirlian photography is a high voltage contact print of an object. The device outlined here provides the high voltage source needed to produce the photographs. The circuit is pretty simple, so you can use point-to-point wiring.

The high voltage transformer used in the circuit is an autotransformer that has three electrical leads. The two wires on the side is where power is supplied by the circuit. The center terminal, the green insulated wire, is the high voltage terminal. Strip about 1 1/2 inches of insulation from this wire to connect it to the exposure plate. **Figure 18-2** shows the inside of the device.

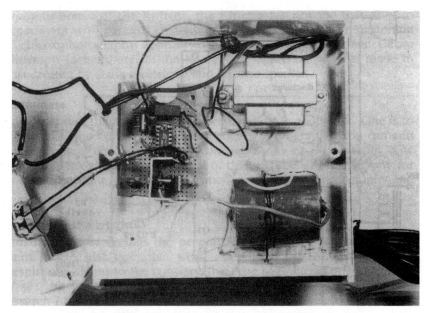

Figure 18-2. The internal view of a Kirlian device.

The housing can be any nonmetallic case made of plastic or wood large enough to house the components. Use a panel mounted normally open momentary contact switch to control exposure. The exposure plate and the controls are placed on top of the enclosure, see **Figure 18-3** and **Figure 18-4.**.

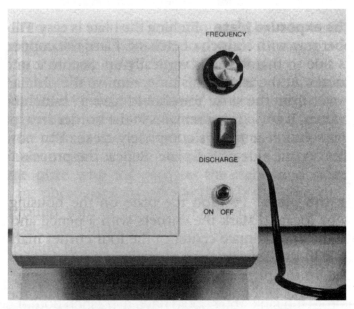

Figure 18-3. The outside enclosure of the Kirlian device.

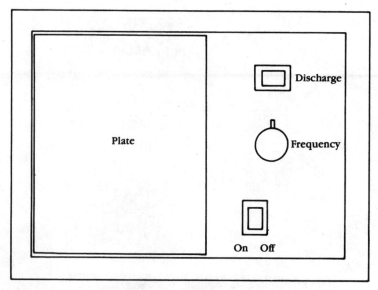

Figure18-4. A Kirlian device.

EXPOSURE PLATE

The plate is the next most important item after the circuit. The plate is constructed from a 4 x 5 inch single side copper board. A 1/2-1 inch border is stripped of copper, using ferric chloride etchant available from Radio Shack.

The border helps prevent electrical arcing from the bottom side (copper side) of the board to the top when making photographs. This also makes it safer to photograph human subjects. You can purchase the plate with a stripped border, or you may make the plate yourself.

MAKING THE EXPOSURE PLATE

Etching the plate is easy. Fill a small plastic or rubber tray with 1/2 inch of etchant. Place the copper plate in the tray on its side, so that it stands vertically up. Secure it in this position for 20 minutes. At the end of this time remove the plate and rinse in running water from the sink. You should have a 1/2 inch border clean of copper.

If any copper remains in the boarder area replace the board in the solution until it is completely clean. You now have a 1/2 inch border on one side of the plate. Repeat the process for the other three sides.

MOUNTING THE PLATE

Position the plate on the housing where you want it to be located. Mark the corners with a pencil and remove the plate. Find the approximate center of the four corner marks and drill a 1/4 inch hole in the housing.

Reposition the plate back onto the top of the housing, using the four corner marks as a guide, copper side down. Mark the location of the drilled 1/4 inch hole on the copper side, using a pencil from the other side of the housing.

Remove the plate and solder a wire in the center of your pencil mark on the copper side. The plate is then secured to the top of the housing permanently. Position the plate copper side down, with the wire going through the drilled hole in the housing. Glue it into place using epoxy or hot glue. When the glue has dried, connect the end of the wire from the copper plate to the high voltage wire on the transformer.

EXPOSURES

What you are photographing determines whether or not the object should be grounded. Grounding an object intensifies the discharge. However, you should only ground inanimate objects. If you are photographing a living subject such as yourself or a pet, under no circumstances should that subject be grounded or be allowed to touch a ground during exposure, as this will lead to a nasty shock.

When photographing an inanimate object such as a leaf, coin, or keys, connect the object to a ground to get a better picture. You do this by connecting a wire to an earth ground, such as a water pipe, then to the object you are photographing.

In a pinch, you can eliminate the ground wire and just touch the object with your finger during exposure.

Exposures are usually made in complete darkness, unless you are using specially packaged film. If you use unpackaged film, you may let a tiny amount of light in, just enough to see what you're doing. I have done this as I shot a few Kirlian photos; so far this light hasn't fogged the film.

Allow a minute or two for your eyes to become accustomed to the darkness. Place the film emulsion side up on the exposure plate (the emulsion is the shiny black side of the film). The film usually has a nick or mark in one of its corners. If you place the nick on your right hand side the emulsion side of the film will be facing you. Put the film on the discharge plate with the nick on the right hand side. Place the object you are photographing on the film. If the object is inanimate, connect a ground wire to it. Turn on the device and press the discharge button for 5 seconds, making the exposure.

Exposure time is determined by trial and error. Start with 5 to 10 seconds, and adjust accordingly. The frequency of the discharge can also be varied to get different discharge effects.

FILM

You can use just about any kind of paper or film to shoot Kirlian photographs. I recommend using 4 x 5 inch transparency film, either tungsten balanced or daylight (see **Figure 18-5**). Both types of film give striking color transparencies. Tungsten balanced film gives colors ranging in the yellows, oranges and reds. Daylight film colors range mostly in the greens and blues.

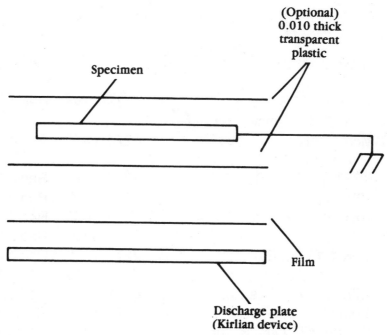

Figure 18-5. A specimen placement on a discharge plate.

SHIELDED FILM

Individually wrapped 4 x 5 inch sheet film is available for shooting Kirlian photographs in daylight. The film is enclosed in a seal black plastic. A notch in the plastic is kept on the right hand side when placed on the device's discharge plate. This orientates the emulsion side of the film facing up. The object is placed on the film and exposed as before. The film is brought to a photolab where the film is removed from the plastic under light-tight conditions and developed.

PARTS LIST

	ITEM		SOURCE
TR1	120 VAC/25.2 VAC 450 mA Transformer	#273-1366	Radio Shack
SW1	Momentary Contact Normally Open	#275-1571	Radio Shack
SW2	SPST 120 VAC Switch	#275-651	Radio Shack
RECT	50V 1 Amp Rectifier	#276-1185	Radio Shack
C1	1000 µF 35V Capacitor	#272-1032	Radio Shack
C2	.022 µF 50V	#272-1066	Radio Shack
R1	2.2K 1/4 Watt Resistor	#271-1325	Radio Shack
R2	50K Potentiometer	#271-1716	Radio Shack
R3	4.7K 1/4 Watt Resistor	#271-1330	Radio Shack
R4	8 ohm 20 Watt Resistor	#271-120	Radio Shack
Q1	TIP 120 NPN Darlington	#276-2068	Radio Shack
IC1	7805 Voltage Regulator	#276-1770	Radio Shack
IC2	4049 Integrated Circuit	#276-2449	Radio Shack
TR2	HV Transformer		Images Company
Q1	IRF830 MOSFET		Images Company
D1	1N5388 200V Zener		Images Company
PC Board			Images Company
4 x 5 Individually Package Color Transparency Film For Daylight Kirlian Photography			Images Company
4 x 5 Copper Clad Board Stripped with 3/8" Border			Images Company

CHAPTER

PINHOLE
PHOTOGRAPHY

Pinhole photography is a fascinating subject. The idea that a small hole, a pinhole in fact, can project an image onto a plane is amazing, see **Figure 19-1**. The cost of making a pinhole camera is minimal, a few dollars at most depending upon your resourcefulness. The camera described here was built for less than one dollar.

HISTORY OF PINHOLE OPTICS

Who was the first to discover that a pinhole could project an image? Photographic folklore is a little fuzzy; some attributed the discovery to the Arabs, others to the Greeks.

Leonardo Da Vinci, however, described the principals of pinhole optics and made sketches for a "camera obscura". The word obscura literally translated means darkroom. The camera obscura is a darkroom with a pinhole that projects a up-

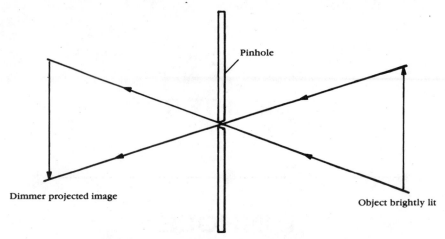

Figure 19-1. A pinhole image formation.

side-down view of a scene outside the room. Artists could trace these projected images onto paper or canvas and use the sketches for paintings. This was before photosensitive materials were discovered to create pictures.

During the Sixteenth Century, portable camera obscuras were used by artists to capture scenes at different locations. The Seventeenth Century Dutch painter Vermeer is believed to have used a camera obscura for many of his paintings.

In 1568 a quantum leap in camera obscura technology was instituted when Danielo Babarbo of Padua suggested using a lens in place of the pinhole. Thus, the outline for a modern camera was complete.

PINHOLE CAMERA

A pinhole camera can be made from just about any light-tight box or can. The size of the box determines the size of film or paper you can use inside it. I built a camera out of a discarded circular cookie can. We'll use this as an example. Just remember you can substitute any other light-tight box or can to your liking.

The circular can measures 7 inches in diameter and 3 inches in height, see **Figure 19-2**. Drill a 1/4 inch hole in the side where you will be placing the pinhole. Paint the inside of the box flat black to prevent any internal reflections. Paint the outside of the box white. The outside is painted white to prevent heat building up inside the can when you are in a sunny location.

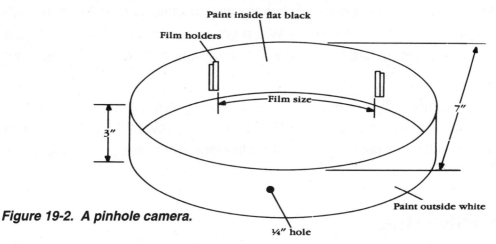

Figure 19-2. A pinhole camera.

On the side opposite the 1/4 inch hole glue two film holders, see **Figure 19-2** and **Figure 19-3**. The film holders can be fashioned from any suitable material: cardboard, plastic or wood. I used two wood popsicle sticks cut in half. Locate the film holders by first drawing an imaginary line from the 1/4 inch hole directly across the can or box. This is the center point of the film holder.

Use a piece of film or paper that you plan to use in the pinhole camera to mark the placement of the film holders. I used 5 X 7

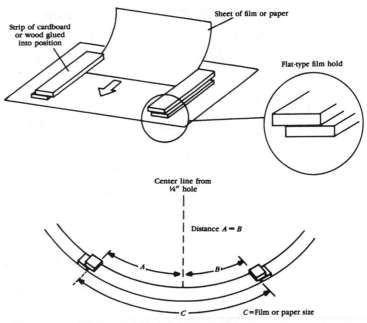

Figure 19-3. Making a film holder.

B/W paper in this camera, cut in half to 2.5 x 7 inches. I removed one sheet of film from the box, (see darkroom, don't open paper in daylight) and used this one sheet of paper to mark where the film holders should go. Hold the paper at its center point in line with the imaginary line from the 1/4 inch hole. Mark the can at each end of the paper. You could of course measure the distance from the center point to each side. However this is an easy way to make a mistake, especially if you're working on a curved surface. When you're finished marking the box, discard the paper, since it has been exposed to daylight and is no longer any good.

Glue in the bottom strip of the film holder. Move them out slightly, making the holder a little larger than what you marked. This will prevent the film from bowing outward when film is placed inside. When the glue dries, secure the top strips.

THE PINHOLE

The pinhole is the most critical piece of the camera. We make our pinhole using a standard hand sewing needle. Needles come in different sizes (different diameters). The following list compares the needle number to its diameter and the optimum focal length for that particular diameter. The focal length corresponds to the distance from the pinhole to the film or paper.

NEEDLE NUMBER	PINHOLE DIAMETER	FOCAL LENGTH INCHES	APPROXIMATE F-STOP INCHES
14	.012	3	250
10	.018	4.5	250
9	.021	6	280
8	.024	8	325
7	.027	10	375
6	.030	12	400
5	.033	14	425
4	.036	16	450
3	.039	20	500

This information is to be used as a guide; it is not written in stone. For instance, I used a .018 inch pinhole with a focal length of 7 inches.

f-Stop

There's an easy way to figure out the f-stop for any pinhole and focal length. f-stop = focal length / pinhole diameter. For my camera this works out to 7/.018 = 388, so the f-stop equals 388. Why would you want to know the f-stop of your camera? It helps figure out how long you should expose your film. But we will go over this later.

Making Your Pinhole

To make your pinhole you need a needle, a piece of thin aluminum fine sandpaper and a piece of cardboard. If you don't have one of the needles specified, use the smallest sewing needle you have. It'll probably work.

You could use a piece of aluminum foil, but it's a little too thin. It's better to use a piece of aluminum from a disposable baking pan available in the housewares department of your local supermarket. A piece of steel or brass shim stock is also a good choice if that's available; .003 is a good thickness.

Cut a 1 inch square of your aluminum or shim stock. Place it on top of the cardboard or paper. Push the needle through the metal sheet about 1/8 of an inch, and rotate the needle, see **Figure 19-4**.

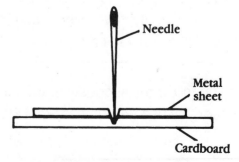

Figure 19-4. Making a pinhole.

Remove the needle and turn the metal sheet over. Take the fine sandpaper (600 Wet/Dry) and gently sand the surface to remove any burrs. Then gently reinsert the needle to remove any debris. Hold the pinhole up to the light to examine it.

Fɪɴɪsʜɪɴɢ ᴛʜᴇ Cᴀᴍᴇʀᴀ

Tape the pinhole on the inside of the can or box so that it lies in the center of the 1/4 inch hole you drilled. The shutter is very simple. If you are using a metal can, as I am, place a small magnet over the 1/4 inch hole. This is your shutter. If you are using another material, tape a black piece of construction paper over the 1/4 inch hole to form a shutter. Your camera's finished. Before we start taking pictures, though, let's get a small pinhole darkroom set up.

Pɪɴʜᴏʟᴇ Dᴀʀᴋʀᴏᴏᴍ

In a pinhole darkroom you don't need an enlarger. What you do need is an area that is relatively light-tight. In many cases a home bathroom or closet can be used. Just stuff a towel by the bottom of the door to block any light from entering.

I recommend using paper in your camera instead of film. With paper there are safelights available that will not fog the paper. This allows light for you to work in your darkroom, for developing, or to load and unload the camera. And as in the case of my camera I needed to cut the 5 x 7 inch paper in half to 2.5 x 7 inches. This task would have been much more difficult if I had to work in complete darkness. With film you have no choice; you must work in complete darkness.

If you follow my suggestion, the second item you need is a safelight. The type of paper you use determines what safelight you need. I recommend you use an RC poly-contrast Kodak B/W paper. For this paper an OC (light amber) filtered safe-light is recommended. (In a pinch try a 7 watt red bulb or red cellophane wrapped around the bulb.)

LOADING THE CAMERA

Under the safelight, remove a sheet of photographic B/W paper from its box. Examine the paper. One side is glossy. This is the emulsion side. The other is a flat white, possibly with the manufacturer's print on it. Load the camera with the emulsion side of the paper facing toward the pinhole.

EXPOSURE

Exposure is determine by four factors: intensity of light, aperture (f-stop), time (shutter speed) and emulsion sensitivity.

Light. Shoot only during daytime using sunlight to begin. Later as you gain experience you may want to try indoor shooting, but this will probably require you to change to film instead of using paper.

Aperture. This is the simple f-stop calculation used when making the pinhole.

Emulsion. The paper emulsions we are using typically have an ISO 4 speed. Films are much faster, up to ISO 3200, which cuts exposure time considerably.

Time. This is the shutter speed. If you like you may ignore all calculations and make a 2 minute exposure. When developing the paper, you can determine if you should increase or decrease your shutter speed.

Figure 19-5 illustrates the aperture opening as compared to the f-stop number. The larger the f-stop the smaller the opening. Each progression of the f-stop number reduces the light intensity reaching the film by 1/2. Most 35MM cameras today either have auto-exposure or in-camera light metering. We will use a trick that was devised for use in case of light meter failure. For taking a bright sunlit picture set your aperture to f/16 (f-stop 16) and your shutter speed equal to the ISO rating of your film. So if you were shooting ISO 200 film your shutter speed would be 1/200 of a second. Now if you increased the f-stop to f/22 you would decrease the light intensity by 1/2 half. To get the correct exposure you would need to double the exposure time and set the shutter speed to 1/100 second. This is the procedure we shall use.

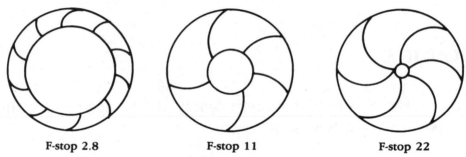

F-stop 2.8   F-stop 11   F-stop 22

Figure 19-5. f-stop and aperature size.

We are using paper that has the approximate ISO speed of 4. So at f/16 we need a 1/4 of a second exposure. Pinhole cameras have pretty high f-stop numbers. The calculation for my camera is f/388. Looking at the following information we see this f-stop is pretty close to the f/360 at 128 seconds. So my exposure time for this particular camera is 2 minutes. Use the calculated f-stop for your camera and the following information to approximate an exposure.

F-STOP	EXPOSURE TIME	F-STOP	EXPOSURE TIME	F-STOP	EXPOSURE TIME
2.8	1/128	16	1/4	90	8
4	1/64	22	1/2	128	16
5.6	1/32	32	1	180	32
8	1/16	45	2	256	64
11	1/8	64	4	360	128
				512	256

SIMPLE DEVELOPMENT

If you are working with paper you can develop the paper using the safelight. You need four trays large enough to hold the paper you are using, see **Figure 19-6**. The temperature required for the chemicals is very forgiving. Room temperature is fine; anything between 65 and 80°F will do.

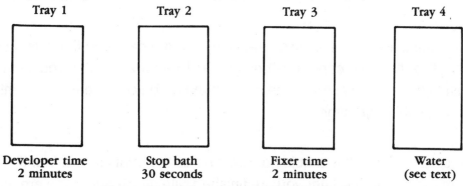

Tray 1	Tray 2	Tray 3	Tray 4
Developer time 2 minutes	Stop bath 30 seconds	Fixer time 2 minutes	Water (see text)

Figure 19-6. The procedure for developing.

Tray 1 contains a developer. You can use just about any development. Mix according to directions on the package. Most developers require about 2 minutes.

You can gauge your exposure by how the picture looks after development. If the paper has a light or faded image it is underexposed, meaning it needs a longer exposure in the camera. If the paper turns very dark or completely black it is overexposed (too much light). Cut the exposure time in the camera.

You can compensate for under- or overexposures somewhat by varying the time in the developer. For underexposed paper, keep it in the developer another 2 minutes or until it reaches a density you like. After about four minutes the developer isn't going to do much more. With overexposed paper, if the paper is turning black, remove it immediately and put it into the stop bath, as this may salvage the print. Proceed with the balance of development.

Although you may salvage some prints this way, it's a good idea to learn from the experience and vary your exposure times in the camera accordingly.

Tray 2 contains a stop bath. This, as the name implies, stops further development. You can buy a commercial stop bath or use plain water. A stop bath is better, and it help prevent the fixer from being contaminated with developer solution. The 30 second time is a minimum, but you could leave the print in the stop bath longer.

Tray 3 contains the fixer. This fixes the image on the paper so that it is no longer light sensitive. After the paper has been in the fixer for 2 minutes you can turn on a regular light. The two minute time is a minimum. You could leave the print in the fixer for a much longer time.

Tray 4 contains water. It is a holding tray. It holds the prints in water until you can bring it to running water. After you are finished with all the prints you are developing, place the tray under running water for 30 minutes or so. This removes the fixer from the paper. If the fixer isn't removed it will eventually turn the print brown and fade the image.

THE PAPER NEGATIVE

If you just finished processing a picture from your camera you may be surprised to find it's a negative. You may be a little wary at this point wondering how good a picture you can get from a paper negative. I know I was, but rest assured the results are very good. You will not be able to tell that the picture was made from a paper negative and not a film negative.

After processing the paper, you can immediately make a contact print before you store the developing chemistry.

If the paper negative you just developed is still wet, that's OK. Under safelight illumination, remove a fresh sheet of paper from its box. Wet this paper with water. Place the papers together, emulsion to emulsion with the negative on top. Put the papers down on a flat surface. Turn on the room light for 1 second. Develop the fresh sheet of paper and you will have a positive print, see **Figure 19-7** and **Figure 19-8**.

Figure 19-7. A paper negative.

Figure 19-8. A paper positive.

You could also do this with dry paper. If so, put a clean piece of glass on top of the sandwich to hold the papers flat.

Always make sure the negative is on top, or you won't get a picture. You can vary the one second exposure depending if the positive is under or overexposed.

GOING FURTHER

We restricted ourselves to using paper in the pinhole camera. After you have gained some experience you may want to try your hand using film. Color transparency and print as well as B/W sheet film is available in 4 x 5 inch size. The disadvantage to using film is that it must be loaded and unloaded in the camera in total darkness. The same is true for development.

The advantages are the ability to take color pictures using much shorter exposures. As a guide, when you double the ISO speed of the film you decrease the exposure time necessary by 1/2. The following information will put you in the ballpark in regard to exposure times for various film speeds.

ISO	MULTIPLY EXPOSURE TIME
4	1
8	.5
16	.25
32	.12
64	.06
125	.03
200	.015
400	.008
800	.004
1600	.002
3200	.001

As an example, for the camera illustrated here, if the camera were load with ISO 64 film we would multiply our 128 second exposure by .06. So 128 x .06 = 7.68 seconds. We'd round that off to 8 seconds and make an exposure.

CHAPTER 20

OIL AND GAS
FROM COAL

The days of coal-driven locomotives and seafaring ships have passed. Even so, coal can still be exploited to provide fuel for today's contemporary cars and home heating systems. Coal can be converted into synthetic gasoline and oil.

The principles of this technology have been in existence for quite a few years. In fact the conversion of coal into an illuminating, cooking and heating gas was an established commercial technology here in the United States in the 1820's. With the advent of the cheaper and greater heating value of natural gas, coal gas quickly disappeared.

The heating value of coal gas ranges from 125 to 560 British Thermal Units (BTU) per cubic foot, depending upon the grade of the coal used and the temperature. Natural gas has a heating value of 1,030 BTU.

The gasification of coal produces a gas that is a mixture of methane, hydrogen and carbon monoxide.

SYNTHETIC GASOLINE & OIL

The conversion of coal into oil and synthetic gas has not been pursued in the United States on any large scale, although large scale operations existed in Germany during WWII. The Germans produced 12,000 barrels of synthetic gas per day from 600 tons of coal.

The conversion of coal into oil and gas requires the adding of hydrogen to the coal. The ratio of hydrogen to carbon in coal is approximately .8 to 1. In oil this ratio is 1.75 to 1.

In large syn-gas plants, water can be used as a source of hydrogen, in the form of high temperature high pressure steam. To make this process economical, the energy to superheat the water to derive the hydrogen from steam must be supplied from coal.

MAKING FUEL GAS FROM COAL

There are numerous ways to generate gas from coal. We will use the simplest method, non-destructive distillation, by heating the coal in the absence of oxygen (air).

Figure 20-1 details our simple fuel gas generator. The reaction vessel is a test tube with a rubber stopper. A short section of glass tubing goes through the stopper to a simple air valve used in home aquariums. The valve is kept open during gas generation allowing the gas to collect in a bag. When gas generation is complete, the valve is closed and the gas generator is removed from the valve and replaced with a simple burner fashioned out of glass or brass tubing, see **Figure 20-2**. With the burner connected, open the air valve and apply gentle pressure on the gas collection bag. Hold a lit match by the burner tip to ignite the escaping fuel gas.

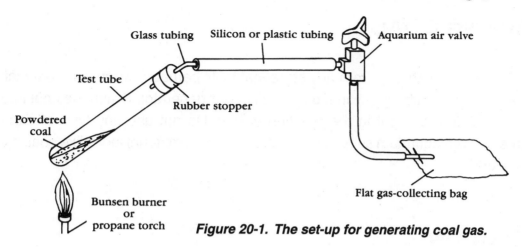

Figure 20-1. The set-up for generating coal gas.

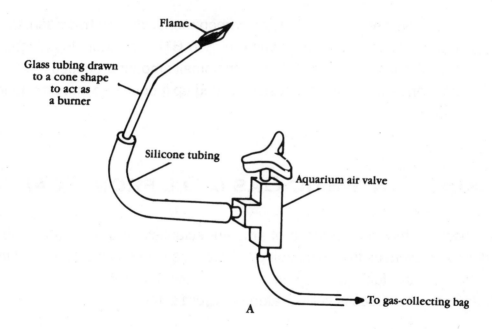

Figure 20-2. The set-up for burning coal gas.

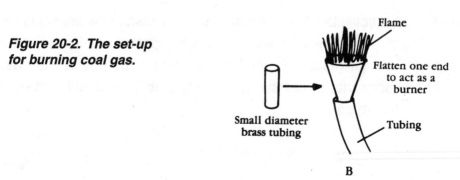

BUYING COAL

Coal can be purchased from any store that sells barbecue supplies. Coal that is intended for a barbecue works quite well in our generator. Crush the coal into a powder before using it in the reaction vessel. Do not use any type of carbon material from aquarium supply stores. Although this material looks like coal, it will not work.

CAUTION

Coal gas, as explained earlier, contains carbon monoxide and hydrogen. Both of these components have a heating value of 300 BTU per cubic foot. Unburned carbon monoxide is poisonous. Although the small amount produced in our generator is not dangerous, if the process is scaled up it does become an important consideration.

MAKING SYNTHETIC GAS & OIL FROM COAL

This process is beyond the reach of amateur scientists at this point in time, but this doesn't prevent us from examining the process. The residue that is left in the test tube from the fuel gas experiment is the raw material needed for making synthetic gas. The process is illustrated in **Figure 20-3**.

The tar residue is collected in a second reaction vessel. This vessel is heated to 1000°F and hydrogen gas is pumped in at 600 lbs. per square inch. The hydrogen combines with the tar residue forming a light motor fuel (synthetic gas) and a heavy fuel oil vapor that is led away from the reaction vessel and condensed using fraction distillation.

◆

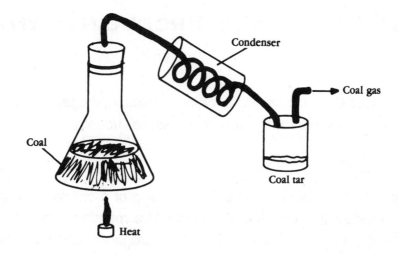

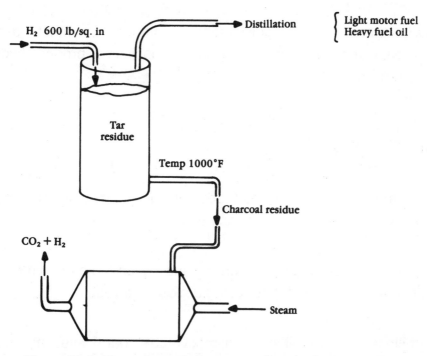

Figure 20-3. The schematic for generating synthetic gasoline.

The residue left from this process is a char residue. The char residue can be utilized to make more hydrogen to drive the primary process.

The above is the simplest method of producing synthetic gas and oil from coal, but it's the least cost effective.

SECOND METHOD FOR PRODUCING SYNTHETIC GAS

The process used by the Germans in WWII is called hydrogenation. It is a method of reacting coal with hydrogen at high pressure, usually in the presence of a catalyst.

The coal is fed into a reaction vessel in the form of a slurry. A catalyst such as cobalt molybdenum is mixed into the slurry. The reaction vessel is heated to a temperature of 850°F and pressurized with hydrogen at 2000 to 4000 psi.

The liquid fraction of the product is distilled to produce a synthetic gas and oil. Unreacted coal is removed from the vessel and gasified to produce hydrogen for the primary reaction.

GOING FURTHER

There are at least two additional methodologies for manufacturing syn-gas from coal; each has it advantages and disadvantages.

There are also processes that can convert and enrich the carbon monoxide and hydrogen to methane (natural gas) using a nickel catalyst.

The United States has two trillion tons of recoverable coal. If an economical method of synthetic gas manufacturing could be developed, it would go a long way in helping the United States recover from the high cost of imported crude oil from the Arab countries.

CHAPTER

◆

ALCOHOL
FUEL PRODUCTION

In the early 1970's our country was subjected to an embargo instituted by the league of Arab nations (OPEC) for our supply of oil. This was a rude awakening for all Americans, who saw our country's energy supply held ransom until the price of crude oil skyrocketed.

The high cost of crude oil made alternate energy sources more feasible. One such avenue we'll take a look at is alcohol fuel production. Although the price of crude oil is not at the point where it is economically profitable to convert the family car to alcohol, with the information contained here you always have the option.

The conversion of sugar into alcohol is a simple process, dating back thousands of years. Yeasts are unicellular fungi that can convert dissolved sugar into ethanol and CO_2. The alcohol and carbon dioxide are byproducts of the yeasts' digestion of sugar. Yeast secretes an enzyme, "zymase", that enables it to digest sugar and promote the chemical reaction. The equation for ethanol production is as follows.

$$C_6H_{12}O_6 \quad \text{Zymase} = \quad 2C_2H_5OH + CO_2$$
glucose   ethanol   carbon dioxide

Yeast is used both in baking as well as in the manufacturing of alcoholic beverages. For baking purposes, the yeast has been bred for its carbon dioxide production, which causes baked goods to rise. This type of yeast is called Baker's yeast. For alcohol production the yeast has been bred for greater alcohol production; consequently, this yeast is called Brewer's yeast.

For our demonstration purposes we can use Baker's yeast that is available in most supermarkets. Although our alcohol yield will not be as high as if we used Brewer's yeast, the ease and simplicity with which we can purchase Baker's yeast makes up for the loss.

SUGAR CONCENTRATION, pH & TEMPERATURE

Sugar concentration is important, as this is the food source for our yeast. A 16% sugar solution is considered the ideal for yeast. A 9% sugar concentration doesn't provide sufficient food, while a 25% concentration is too much food, which would initiate the yeast to put more energy into reproduction instead of producing alcohol.

The ideal pH of the solution should be 4.5 to 5.0 . Ideal temperature ranges from 75°F to 85°F. We are going to fudge from the ideal yeast environment. It would cost too much in time, labor and money to provide the ideal environment. You would only meet this criteria if you were pursuing alcohol production on a larger basis.

REACTION VESSEL

Your reaction vessel can be any small jar with a tight fitting lid. The design is simple, see **Figure 21-1**. The tubing and fittings are available from a local tropical fish store. The tubing you should use is a silicone base (light transparent green) rather than plastic (clear). The reason to use silicone is that the tubing is more flexible.

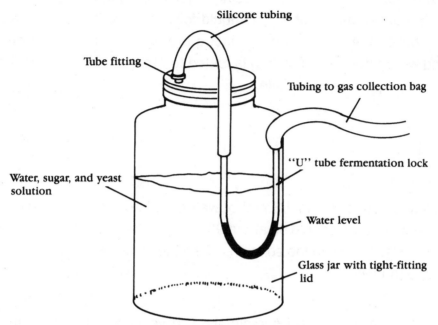

Figure 21-1. The reaction vessel.

FERMENTATION LOCK

On the vessel we have a simple fermentation lock. This is the U-shaped tube partially filled with water. The function of the lock is too allow carbon dioxide to escape but at the same time prevent oxygen (air) and unwanted organisms from entering the reaction vessel. If air (oxygen) is allowed to enter during fermentation the yeast will convert the sugar into vinegar (acetic acid) instead of alcohol.

Use a larger diameter tubing than is shown in the illustration or the CO_2 will push the water through the tubing.

The fermentation lock is essential to the production of alcohol and also prevents the carbon dioxide from building up to a hazardous pressure.

When running my experiment I placed a gas collecting bag on the fermentation lock. I did this for two reasons: First, to measure the carbon dioxide output of the process; Second, it showed fermentation was occurring, and pinpointed when it stopped (no more gas production). Using a gas collecting bag is not essential to carry out the experiment, but if you should follow my experiment, note I used a bag, not a balloon. A balloon would require pressure to inflate, and pressure is something we do not want. The bag is folded to remove as much air as possible and requires no pressure to inflate.

STEP 1

Mix a tablespoon of sugar in 1/3 pint of water, and add into the reaction vessel. The water temperature should be about 90°F to get the yeast started. Add a small quantity of Baker's yeast to the solution. Close the vessel and attach the tubing to the fermentation lock.

Do not fill the reaction vessel completely. The yeast mixture will foam when it starts production, so if the vessel is completely filled the foam will be forced through the tubing.

FERMENTATION

Fermentation develops quickly. The process is completed in about three days. However, because the yeast has finished working doesn't mean all the sugar has been converted. The alcohol is toxic to the yeast. Fermentation stops for Baker's yeast when the alcohol concentration reaches 8-12%. For Brewer's yeast fermentation continues a little longer, until the alcohol concentration reaches 10-14%.

DISTILLATION

It is necessary to distill (remove) the alcohol out of the water for it to be useful as a fuel (see **Figure 21-2**). When fermentation has stopped, it's a good idea to distill the alcohol/water mix as soon as possible. If the mix becomes contaminated with air it can still turn the alcohol to vinegar.

Alcohol is hygroscopic, meaning it readily combines with water. Water boils at 212°F, ethanol at 173°F. You may think that by raising the temperature of the solution to 175°F you could boil off (remove) all the alcohol, but unfortunately this will not happen. The water and alcohol molecules combine tightly, and the heat of vaporization requires additional energy. With all things considered we should bring the solution to about 200°F. The majority of the steam will be alcohol and will yield a 40-50% concentration. This alcohol concentration is good enough for our purposes and can be used as a fuel for an alcohol lamp.

Remember, if you used a tablespoon of sugar in the reaction vessel, don't expect to distill more than a tablespoon of alcohol.

To increase the concentration of alcohol further requires additional distillation steps. Running the finished product from the first distillation through the distillation process at a lower temperature of 190°F will increase the alcohol concentration to about 65%. Repeating the process again at 180°F will increase the concentration to about 90% (Fuel Grade).

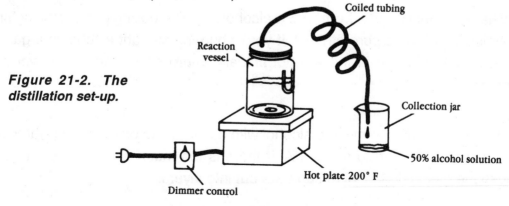

Figure 21-2. The distillation set-up.

STEP 2

When fermentation has stopped (in about three days), we must distill the alcohol out of the solution. For this you need a hot plate and dimmer control. Hot plates are available in most houseware and hardware stores, as are dimmer controls. Dimmers are usually used for incandescent lighting. Just be sure the capacity of the dimmer (in watts) equals or exceeds the wattage of the hot plate. Wire the dimmer in the hot plate's line cord. Use an oven thermometer to measure the temperature of the hot plate. Use the dimmer control to calibrate the hot plate's temperature to about 200°F.

Another method of calibration is to place a cup of water on the hot plate with the thermometer in the water. Use the dimmer control to adjust the water temperature to 200°F.

With the hot plate calibrated we can start distilling the solution. Remove the tubing going to the fermentation lock at the top of the reaction vessel. Replace this with a longer coiled section of tubing that leads into a collection jar. The liquid that collects should be about 50% alcohol, which is good enough to use in alcohol lamps.

FUEL GRADE

It is hard to remove water from a 95% alcohol mix. The boiling point of the solution is almost identical to pure ethanol. If you're burning straight alcohol in a gas engine, 5 to 10% water in the alcohol actually improves performance over pure alcohol.

However, if you plan to mix the alcohol with gas to make gasohol, the water content must be reduced to about 1%. If there is too much water in the gas it will cause the gas and alcohol to separate out into layers.

◆

USING FUEL GRADE ALCOHOL

Running fuel grade (90-95%) alcohol in a car is probably easier than you think. Simply enlarge the carburetor jets by 30% and advance the engine timing several degrees and you're ready to go. A little extra equipment is necessary for cool weather starts (below 50°F), such as priming the engine with a little gas to start it up. A preheater or a heat pickup off the exhaust manifold will provide warm air for proper alcohol vaporization to keep the engine running smoothly.

The use of ethanol fuels in cars is not new. Henry Ford offered automobiles at the turn of the century capable of running on either gas or alcohol.

2-CYCLE ENGINES

If you want to use alcohol to run smaller 2-cycle gas engines for lawn mowers and such, it is necessary to mix a vegetable oil in the fuel instead of motor fuel that is used with gas. The reason is that motor oil doesn't mix as well with alcohol as it does with gasoline.

U.S. GOVERNMENT REGULATIONS

The U.S. government requires that all stills producing alcohol be registered. The Bureau of Alcohol, Tobacco, and Firearms have several regional offices across the United States. For your regional office, call or write to:

Department of the Treasury
Bureau of Alcohol, Tobacco and Firearms
Distilled Spirits and Tobacco Branch
650 Massachusetts Ave
Washington, DC 20226
(202) 927-8210

Contact the regional office for an AFP (alcohol fuel permit). Currently the AFP doesn't cost anything, but there is a another fee involved. Mention that your still is a demonstration unit that only produces a small amount of fuel alcohol, like 10 ml per run, and they may waive the fee.

CHAPTER

Bio-Gas
Generation

Waste disposal is a growing problem. Currently most human waste is released into the planet's water supply with perhaps a little sewage treatment. The pollution has negative effects on all levels of the ecosystem.

How much better it would be if we could purify this waste and produce a non-polluting energy source. Well we can, with the added benefit that the residue from the process is a nutrient-rich fertilizer for plants.

Bio-gas generation plants are currently used in India and Australia, most commonly found on rural farms. The "rural" location adds the incentive to produce one's own power (as much as possible) rather than paying to ship it in.

BIO-GAS COMPOSITION

Bio-gas can be generated from human, animal or vegetable waste. The composition of Bio-gas is approximately 50-60% methane, 30-35% carbon dioxide, hydrogen 1-5%, nitrogen .5-3%, with trace amounts of carbon monoxide, oxygen and hydrogen sulfide.

Bio-gas can be used for heating, cooking and as a fuel supply for gasoline engines (with minor modifications).

PLANT FERTILIZER

The residue from the process is a high quality nutrient-rich plant fertilizer. This is an important factor. Utilizing the fertilizer we can create a more closed ecosystem that doesn't waste material and is self-sufficient. **Figure 22-1** shows a closed system.

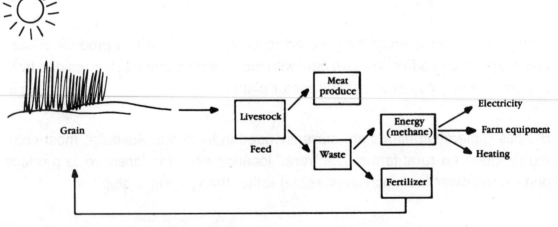

Figure 22-1. The recycling of waste.

ANAEROBIC FERMENTATION

Bio-gas is generated by the decomposition of organic material (manure or human feces and/or vegetable waste) in the absence of oxygen (air). In this respect it is similar to alcohol production. If we allow oxygen into the alcohol fermentation process we produce vinegar instead of alcohol. With Bio-gas generation, oxygen in the process will produce ammonia instead of methane gas.

Vegetable waste produces, on average, 7 times more methane gas than animal waste. Digesters have been designed that work primarily on plant material, but additional nitrogen must be added to the mix.

TEMPERATURE

The ideal temperature for the anaerobic digestion of waste is 85-105°F. Although digestion can still occur from a temperature range of 50 to 120°F, it will proceed much slower.

ACIDITY

The ideal acidity is 6.8 to 8.0 pH.

CARBON TO NITROGEN RATIO

To maximize the process, the ratio should be 30 parts carbon to 1 part nitrogen. The bacteria consume carbon 30 times faster than nitrogen. Having the proper ratio insures complete digestion of the raw materials, providing maximum gas production and a residue (plant fertilizer) with the highest nutrient value.

◆

Animal waste usually has a high nitrogen content while vegetable waste usually has a higher carbon content.

TEST EXPERIMENT

For our small test experiment, we will not concern ourselves with trying to set the ideal environment for anaerobic digestion.

Any small bottle with a tight fitting lid can be used for a digester, see **Figure 22-2**. Mix some manure and water to form a slurry. Add the slurry mixture to the digester. Do not fill the digester to the top as there might be some foaming as the bacteria gets to work. If you wish, add a small amount of sawdust to the slurry to increase the carbon content. The sawdust is optional, as the digester will produce bio-gas without it. However, if you do add sawdust make sure it is from "real" wood and not some particle board or wood composite. Wood composites contain resins and epoxies that may halt gas production. The output of the digester is fed through an aquarium air valve into a gas collecting bag.

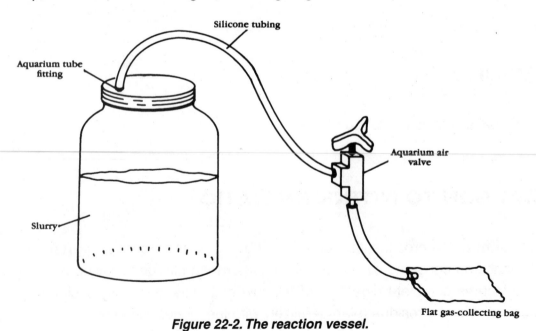

Figure 22-2. The reaction vessel.

The digester should be held at a temperature of 80-90°F. You could use an aquarium heater and a water jacket to keep the digester at this optimum temperature. Or, you may locate the digester in a warm section of your home or apartment, perhaps by a heating vent, furnace or hot water tank.

Gas production usually begins within 21 days. When the unit begins producing gas, it is mostly carbon dioxide. To test for methane production, close the air valve and remove the bag section, keeping the tubing to the bag crimped or folded at this point to prevent any gas from escaping from the bag. Hold a lighted match near the free end of the tubing and release the crimp. Carbon dioxide will not burn. Continue to test the gas production every day or so until the match ignites the escaping gas. Remember to open the air valve after you reconnect the gas collecting bag, or you may find an undesirable cleanup job waiting for you the next time you check on the digester.

When you are producing methane, allow the gas to collect in the bag. You can construct a simple burner (see **Figure 22-3**) to burn your collected gas. When you are burning the methane gas you can make a more spectacular demonstration by squeezing the bag to increase gas pressure.

When your digester has stopped producing gas, the residue left in the bottle is an excellent fertilizer for your plants.

GOING FURTHER

The most obvious avenue for research is sewage treatment plants. Bio-gas plants cannot only produce their own non-polluting power but are ecologically sound investments.

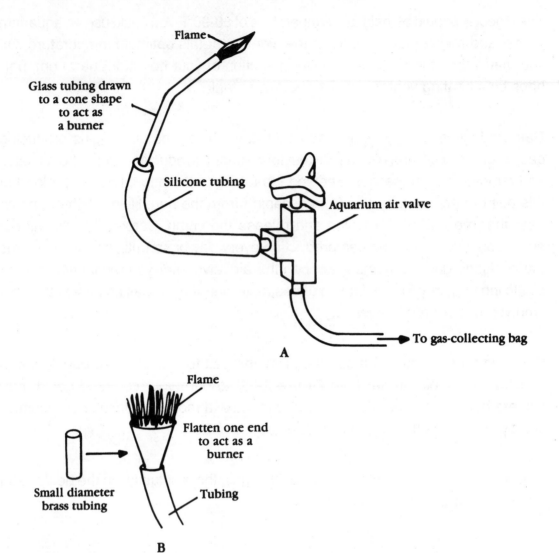

Figure 22-3. The set-up for burning bio-gas.

CHAPTER 23

◆

WOOD AS FUEL

You may think wood is already a fuel. It's common knowledge that there are wood burning stoves and ovens, but did you know that you could run gasoline engines on wood?

The method to extract a motor fuel from wood is simple. You may distill wood alcohol (methanol alcohol) or a gaseous hydrocarbon from wood. Running gasoline-powered vehicles from wood isn't a new idea. During the Second World War, civilians in Europe did so.

Modifications to the carburetor were necessary for the engine to run this type of gaseous fuel.

EXPERIMENTAL GAS GENERATOR

Figure 23-1 illustrates our simple gas generator. The generator is filled with saw-dust or small wood chips. It's important to use real or natural wood material. Particle board and other wood substitutes contain resins and epoxies that may develop harmful fumes.

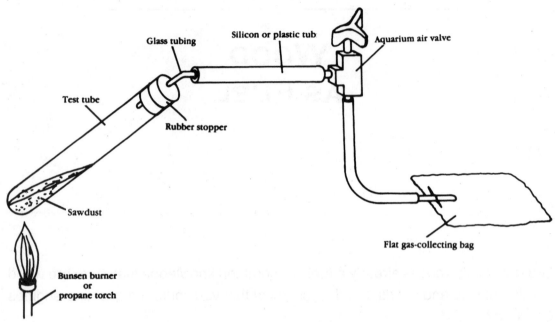

Figure 23-1. The set-up for generating fuel gas.

The sawdust is heated in the enclosed test tube. The gas generated is fed into a gas collecting bag. When the wood is completely utilized, close the air valve and disconnect the generator. Replace the generator with a gas burner, see **Figure 23-2**. Open the valve and use a lighted match to ignite the escaping gases.

The gas composition is a mixture of carbon monoxide and hydrogen, with a small amount of methane and carbon dioxide. The residue left is mostly carbon and may be utilized by another process, such as a carbon source for the Bio-gas generator.

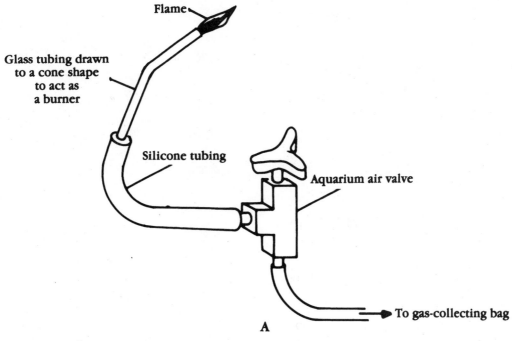

Flame

Glass tubing drawn
to a cone shape
to act as
a burner

Silicone tubing

Aquarium air valve

To gas-collecting bag

A

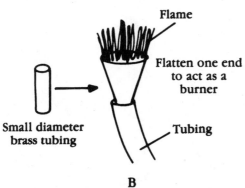

Flame

Flatten one end
to act as a
burner

Small diameter
brass tubing

Tubing

B

Figure 23-2. The set-up for burning fuel gas.

APPENDIX A
SUPPLIERS INDEX

◆

ALLEGRO ELECTRONICS SYSTEMS
3 Mine Mountain Road
Cornwall Bridge, CT 06754
(203) 672-0123

EDMUND SCIENTIFIC
101 E. Gloucester Pike
Barrington, NJ 08007-1380
(609) 573-6250

ELECTRIC FIELD MEASUREMENT
PO Box 326   Rte. 183
W. Stockbridge, MA 01266
(413) 637-1929

HERMANT CHIKARMANE
Marine Biological Labratory
Woods Hole, MA 02543

IMAGES COMPANY
PO Box 140742
Staten Island, NY 10314-0024
(718) 698-8305

MOUSER ELECTRONICS
PO Box 699
Mansfield, TX 76063
(800) 34-MOUSER

NEWARK ELECTRONICS
4801 N. Ravenswood Ave.
Chicago, IL 60640-4496
(312) 784-5100

THE NUCLEUS COMPANY
601 Oak Ridge Turnpike
Oak Ridge, TN 37830
(615) 483-8405
(800) 255-1978

WORKS CITED

1. *Negative air ion effects on learning disabled and normal achieving children.* Morton LL; Kershner JR. University of Windsor, Faculty of Education, Ontario Canada. 5/90

2. *Effect of negative ion generators in a sick building.* Finnegan MJ; Pickering CA; Gill FS; Aston I; Froese D. Department of Thoracic Medicine, Wythenshawe Hospital, Manchester England. 5/87

3. *Aeroionization in prophylaxis & treatment of respiratory in calves.* Sologub, T.I.; Borzenko, N.F.; Zemlyanskiy, V.P.; Plakhotnyy, K.F. Russia 1984.

4. *Effect of ionization on microbial air pollution in the dental clinic.* Gabby J; Bergerson O; Levi N; Brenner S; Eli I. Research Institute for Environment Heath, Sackler School of Medicine, Tel Aviv, Israel. 6/90

5. *Effects of ionized air on the performance of a vigilance task.* Brown GC; Kirk RE; Systems research Laboratories, Inc., Brooks AFB, TX 6/87.

6. "Negative Ions". Wright JE; Muscle & Fitness Magazine 1/91.

7. *Effect of Artificial Air Ionization on Broilers.* Stoianov P; Petkov G; Baikov BD. Vet Med Nauki 1983 Bulgaria.

8. *55-Nitinol,* C.M. Jackson. NASA Pub. SP-5110, Washington D.C. 1972.

9. "On the Thermodynamics of Thermoelastic Martensitic Transformations". R.J. Salzbrenner and Moris Cohen in Acta Metallurgica. Vol 27, No. 5, May 1979.

10. "Shaped-Memory Alloys". L. McDonald Schetky. Scientific American Vol. 241 11/79.

INDEX